TRACING YOUR
ARMY ANCESTORS

SIMON FOWLER

Pen & Sword
MILITARY

ISBN 1-84415-410-6
ISBN 978-1-84415-410-4

The right of Simon Fowler to be identified as
Author of this Work has been asserted by him in accordance
with the Copyright, Designs and Patents Act 1988.

A CIP catalogue record for this book
is available from the British Library

Typeset in 10/12pt Erhardt by
Concept, Huddersfield

Printed and bound in England by
CPI UK

Pen & Sword Books Ltd incorporates the Imprints of
Pen & Sword Aviation, Pen & Sword Maritime, Pen & Sword Military,
Wharncliffe Local History, Pen and Sword Select,
Pen and Sword Military Classics and Leo Cooper.

For a complete list of Pen & Sword titles please contact
PEN & SWORD BOOKS LIMITED
47 Church Street, Barnsley, South Yorkshire, S70 2AS, England
E-mail: enquiries@pen-and-sword.co.uk
Website: www.pen-and-sword.co.uk

Contents

If I had time to study war, I think I should concentrate almost entirely on the 'actualities of war' – the effects of tiredness, hunger, fear, lack of sleep, weatherit is the actualities that make war so difficult and are usually so neglected by historians.

Field Marshal Lord Wavell to Basil Liddell Hart
(quoted in Ilana Bet-El,
Conscripts: Lost Legions of the Great War, Sutton, 1999)

Preface

The most popular subject for research at The National Archives at Kew is the military records. Every day dozens of researchers pore through the records and peer at microfilm trying to find out about ancestors who served in the British Army. This book is designed to help them and other enthusiasts find their way through the records and provide background on the life of ordinary officers and soldiers.

That researching military ancestors is so popular is perhaps not surprising. In part this is because military history is immensely popular with a multitude of books published each year (particularly on the two world wars), as well as a plethora of television programmes, not forgetting the thousands of websites which have sprung up in recent years. People are fascinated with the brave deeds and the minutiae of warfare. As Dr Johnson wisely said, over 200 years ago: 'Every man thinks meanly of himself for having not been a soldier or not having been at sea.'

Most families have one or two men who served in the Army, generally of course during the two world wars, but surprisingly often at other times. And unlike for almost every other occupation, full historical records survives.

Working class men driven by patriotism or more often the baser sentiments of hunger and unemployment were the mainstay of the British Army, poorly treated in peacetime and regarded as heroes in wartime. There was a saying that Jack Frost was a great recruiting sergeant. Men from impeccable middle class homes occasionally felt the need to join up as privates, but many more played as weekend soldiers in the Militia and volunteers.

Until the First World War few working class men crossed the class divide to become officers. Officers, particularly in cavalry and infantry regiments, came from a narrow landowning and public school educated social class. The Royal Artillery, Royal Engineers and other corps and units were less exclusive and inevitably made up more of professional men.

In the Napoleonic Wars 80 per cent of Wellington's armies were made up of ordinary infantry soldiers. This proportion changed as better and more complex technology was introduced, such as more powerful guns and better communications, and the needs of the soldiers began to be recognised, with proper medical care and three square meals a day. However it was not until the Second World War that more soldiers were engaged in the 'flummery of war' as Churchill once said, rather than at the 'sharp end'.

Our soldier forebears generally lived interesting lives. Even if they did not, having a soldier on the family tree is rather more romantic that the normal agricultural labourers ('ag labs') and framework knitters from whom we are mostly descended.

At the end of his memoirs of the Peninsular Wars, Rifleman Benjamin Harris sums up the experience of the British soldier not just in Spain and Portugal, but perhaps across the centuries:

> *Let me here bear testimony to the courage and endurance of that Army under trials and hardships which few armies, in any age, can have endured. I have seen officers and men hobbling forwards, with ragged backs, without even shoes or stockings on their bleeding feet, and with tears in their eyes from the misery of long miles and empty stomachs (and it took a lot to bring a tear into the eye of a Rifleman of the Peninsula). Youths not long removed from their parents' home and care, officers and men, bore hardships and privations which, in our own more peaceful days, we have little conception of. Yet these same men, though faint and weary with toil, would brighten up in a moment when the word ran amongst us that the enemy were at hand.*
>
> *The field of death and slaughter, the march, the bivouac, and the retreat, are not bad places in which to judge men. Having had the opportunity of doing so, I would say that the British are amongst the most splendid soldiers in the world. Give them fair play, and they are unconquerable.*
>
> *I enjoyed life more whilst on active service than I have ever done since, and I look back upon my time spent in the fields of the Peninsula as the only part worthy of remembrance. As I sit at work in my shop in Richmond Street, Soho, scenes long passed come back upon my mind as if they had taken place but yesterday. I remember the appearance of some of the regiments engaged. And I remember too my comrades, long mouldered to dust, once again performing the acts of heroes.*

Arrangement

Compared with the Royal Navy and, to a lesser extent, the Royal Air Force there are surprisingly full records for the Army. And if one set of documents is missing then it is often possible to find similar information from another source. They are also generally easy to use with few pitfalls for the unwary. As Gerald Hamilton-Edwards points out, this: 'means that a great deal more can be discovered about ancestors who served than about those who remained in civilian life. The [Army] had no idea, of course, when they called for these returns, that they were going to aid future genealogists.'

The vast majority of records are to be found at The National Archives at Kew, generally on microfilm, but you may well get to handle original muster rolls and war diaries. At the time of writing only a small proportion of records have been made available online, but this is likely to change over the next few years.

The book is arranged in seventeen sections which either concentrate on an aspect of life in the Army before 1914 or the wars and conflicts of the twentieth century. If you are a complete beginner it is a good idea to start with Chapter 1, which looks at how to begin and how the basic genealogical sources can help your research.

Subsequent chapters generally comprise an introductory essay providing background and the summary of the major records which may provide clues to your ancestor's career in the Army. By doing so I hope to provide a flavour of what life was like for officers and men and, where appropriate, the constraints that affected their service.

The book covers men who served in the Army when a permanent service was raised in the 1660s (although there is a section on records before then) through the next three centuries until the 1960s, when it becomes increasingly difficult to research because most records of any use are still retained by the Ministry of Defence.

My own familial connections with the Army are not great. A great-uncle of mine was a conscript in the Rifle Brigade during the First World War and was killed ten days before the Armistice south-east of Kortrijk in Belgium. I have been unable to find out very much about him largely because his service record (like millions of others) has not survived. So little impact did Great-Uncle Stanley make on the war that I have come across three different dates of death for him. As far as I can tell no ancestor served in the Army during the nineteenth century (they joined the Navy instead) or in the Second World War (preferring the Air Transport Auxiliary).

That said Army history has always interested me – not so much the battles and campaigns or even the weaponry, all turgid and confusing – but what it was really like for the officers and other ranks. I hope that this comes across here and will inspire readers to find out more for themselves.

Parts of this book have appeared in various forms in *Ancestors Magazine*, *Family History Monthly*, *Army Service Records for Family Historians* (Public Record Office, 1989), and *Tracing your Ancestors in the First World War* and *Tracing your Ancestors in the Second World War*, both published by Countryside Books.

Acknowledgements

Thanks to Rupert Harding and his colleagues at Pen and Sword for their help and support, particularly Susan Econicoff who edited the text.

I also would like to thank Sylvia Levi, who irritatingly but probably sensibly distracted the author by making him go to concerts and otherwise meet people who hadn't been dead for a long time.

Errors and omissions are of course my own.

Chapter 1

GETTING STARTED

1.1 Where to start

The best place to begin your research is to work out what you know already. Write down definite facts as well as anything about which you are not sure.

- The full name of the person you are researching, as well as any variants that you might be aware. Your father, known as Tommy Atkins, might have enlisted as John Thomas Atkins or be in the records as TJ Atkins or even have the name misspelled by the clerks as John Atkens.
- Which regiment or corps he served with.
- When he enlisted (usually at about the age of eighteen) and when he was discharged.
- Date of death if killed in action or died of wounds.

In an ideal world you should also know his:

- Regimental number.
- What injuries or disabilities resulted from his war service – they may have been physical such as an artificial leg or shrapnel in the body or psychological, perhaps plagued by recurring nightmares?
- The medals he was awarded.

This book will help you follow up these leads and give you ideas about where else you might look for information. Once you start, you might be pleasantly surprised about what you can find out. If you get stuck check the problem solving appendix on page 178.

1.2 Background research

It is a good idea to familiarise yourself with the period, and perhaps get the feel for the life experienced by your ancestor, by doing some background reading. To help I have included introductory essays before each chapter which describe in brief the Army in which your ancestor served and some pointers about the life they led. By doing so you will gain some idea about the life that your ancestor lived and the events he experienced. There are a huge numbers of books, particularly for the two world wars. A selection of suggested books can be found in most chapters and there is also short bibliography in Appendix 5.

If you've got internet access you can find out what is currently in print by visiting the Amazon website (www.amazon.co.uk): most bookshops can order any book currently in print for you – surprisingly, this is often cheaper and quicker than ordering via Amazon. There are also a number of booksellers who specialise in new or second hand military history. One of the largest, with an excellent catalogue, is Naval & Military Press, Unit 10, Ridgewood Industrial Park, Uckfield, TN22 5QE, www.naval-military-press.com.

The British Library Public Catalogue (http://blpc.bl.uk) will supply you with details of virtually every book ever published in Britain. You can borrow most books through the inter-library loan service for a few pence, so talk to your local library staff (ideally in the reference or local studies library) to see what they can find for you.

1.2.1 The internet

An increasing amount of information is available online. I have assumed that readers have the internet at home or at work, but if you do not, libraries have computers linked to the web and there is often training available which will show you how to make the best of the net. Website addresses (URLs) were accurate at time of going to press, but if you find any which don't work you should be able to find the answer (and any related sites) by using a search engine such as Google (www.google.com).

You will find addresses of websites scattered through the book. Apart from official sites I have included unofficial ones which I felt were serious and reasonably accurate. There are a number of sites which offer basic information to help you research your military ancestors, although they are not always as up-to-date as they might be. Good introductory pages include those by the Army Museum Ogilby Trust at www.armymuseums.org.uk and the Imperial War Museum (www.iwm.org.uk/server/show/nav.00100a) as well as Fawne Stratford-Devai on 'British Military Records Part 1: The Army' in The Global Gazette at www.globalgenealogy.com/globalgazette/gazfd/gazfd44.htm, and Jay Hall on 'British Military Records for the 18th and 19th Centuries' (www.genuki.org.uk/big/MilitaryRecords.html).

Links to many British Army websites can be found at www.cyndislist.com/miluk.htm#Battles.

Websites may not be much help in providing answers to particular problems once you start researching an individual soldier. Alternatives are the various online discussion forums which are ideal places to ask questions or learn from the experience of others. I belong for example to a First World War mailing list which has a constant stream of requests for help with members' research which are patiently answered by other list members. In addition there are snippets of news and debates on various related topics: a recent one was how it was possible to enlist under a false name!

The best index of mailing lists is on the Genealogy Resources on the Internet site, which has a page devoted to Wars and Military matters at www.rootsweb. com/~jfuller/gen_mail_wars.html. For details of those specifically relevant to the British Isles, visit www.genuki.org.uk/indexes/MailingLists.html. Indeed this site gives almost all genealogical mailing lists, so there are also links for particular surnames.

There are also lists for military history rather than military genealogy which may be of help in understanding historical or regimental background. Yahoo Groups also has a relevant mailing list, Britregiments (http://groups.yahoo. com/group/britregiments). In addition British-Genealogy has a number of historical forums for specific wars and periods of military history, all listed at www.british-genealogy.com/forums.

1.3 Where the records are
Eventually you will have to use original archive material, which may prove to be time-consuming but in return is deeply addictive and rewarding. These documents can be found in one of several places: a national repository or museum, a regimental or service museum, or local studies library and county record office.

Most records of the armed forces are held at national repositories such as the National Archives (formerly the Public Record Office), or museums such as the National Army Museum and, for the twentieth century, the Imperial War Museum.

Other important resources are regimental museums which may have archival material relating to their regiment or service, although their holdings vary greatly. Certainly they should have collections of photographs, the regimental journal and may be able to tell you more about the men (particularly officers) who served in the regiment.

A few regimental archives are now with county record offices: the Durham Light Infantry archive, for example, is at the Durham Record Office and the Manchester Regiment with Tameside Local Studies Library. Regimental museums are described in Terence and Shirley Wise, *A Guide to Military Museums and Other Places of Military Interest* (10th edition, Terence Wise, 2001) or visit www.armymuseums.org.uk. Local record offices and local studies libraries may also have files relating to territorial and Militia units.

You may get started by trying to research the people who appear in uniform in family photographs, such as this First World War officer. (Jane Starkie)

There are also a large number of specialist repositories for businesses, universities and charities that may (just possibly) have information. Company archives, for example, may include papers about employees who served in the forces or the provision of a memorial plaque for those who did not return.

Unfortunately there is no hard or fast rule as to who has what, so the papers of a particular commander may be at a local archive or material about a local militia unit could be at The National Archives. The National Register of Archives (www.nationalarchives.gov.uk/nra) lists many holdings at local record offices, some museums and specialist repositories. An easier to use alternative is the Access to Archives database (www.nationalarchives.gov.uk/a2a). Neither is complete, however, and neither specialises in military subjects.

Every county has a local record office and most towns have a local studies or history centre and they all have websites of varying degrees of usefulness. The ARCHON section of The National Archives website (www.nationalarchives. gov.uk/archon) provides links to local record offices. It can be much harder however to track down local studies libraries, particularly ones outside your area. You can, however, find most addresses at www.familia.org.uk.

Janet Foster and Julia Sheppard's *British Archives: a Guide to Archive Resources in the United Kingdom* (3rd edition, Palgrave, 2001) lists the vast majority of archives in Britain – every reference library should have a copy. Less comprehensive, but more practical, is Christine Morris and Gillian Rayment, *Record Offices: How to Find them* (Federation of Family History Societies, 2006). Addresses (with links to websites where appropriate) can also be found online at www.nationalarchives.gov.uk/archon.

Each record office has a different system of managing its records, although most follow the same principles of archive administration. Documents are kept together by collection, rather than rearranged by subject as happens in a library.

Some Army records have been filmed by the LDS (Mormon) Church, particularly the First World War service records. They have a network of local Family History Centres attached to their chapels, which are open to all free of charge. The Centres will order films from the main Family History Library in Salt Lake City for a small fee. It may be cheaper and more convenient to do this than visit The National Archives at Kew. More details can be found at www.familysearch.org or ring 020-7589 8561 for details of your nearest centre.

1.4 Using the National Archives
The vast majority of records described in this book are to be found at The National Archives (TNA) at Kew (address at end of chapter). Although an increasing proportion of records are available online, you will almost certainly need to visit Kew during the course of your research. Although at first sight daunting, TNA is actually well organised and welcoming and you should be able

to find what you want with the minimum of fuss. Before you start you need a reader's ticket, which is free and lasts three years. You can pre-register online at www.nationalarchives.gov.uk/registration or when you arrive. However you will need to bring with you some form of identity such as driving licence, passport or credit/debit card. There is a very good introductory leaflet for new readers which the receptionist will give you. There is also a simple illustrated step-by-step guide to visiting Kew and tracing a Military Cross for the First World War in the February 2006 (issue 42) of *Ancestors Magazine.*

You are welcome to use a laptop computer in the reading rooms (and there is a wireless connection) and, provided it is registered in the Reprographic Copying Room, a digital camera. You are of course only allowed to take pictures for your own use.

A concise introduction to the holdings is at www.nationalarchives.gov.uk/ familyhistory/military/army, while there is more extensive material on the 'in-depth learning guides' pages at www.nationalarchives.gov.uk/pathways/ familyhistory. This has pages devoted to some of the most important resources available at TNA including: soldiers' papers, tracing Army officers, regimental diaries, war medals, and the British Army before the First World War.

An invaluable source are the dozen or so Research Guides on Army records produced by The National Archives, which simply explain which documents are available for a particular topic at Kew and, occasionally, elsewhere. They are listed in Appendix 3 and at www.catalogue.nationalarchives.gov.uk/ researchguidesindex.asp.

The National Archives also publish a number of guides to family history, including the comprehensive *Tracing Your Ancestors at The National Archives* (8th edition, 2006) by Amanda Bevan. At the time of writing their other military guidebooks were rather out of date but books on medals and a general introduction to researching the three services will be published in 2006.

All of the 9,500,000 or so records at The National Archives are described in the online catalogue. This easy to use resource also includes the names of many soldiers, including Army officers of the First World War and ordinary soldiers between 1760 and 1854 who received a pension. It has recently been upgraded to include all the other databases (most notably the Access to Archives database) available on the TNA website.

TNA records are arranged by department (sometimes called lettercode) – you are most likely to use records created by the War Office (prefix WO). The types of records are arranged by series (or class), so war diaries for the First World War are in series WO 95, Militia Attestation Papers (1806–1915) are in WO 96, and Soldiers Service Documents (1760–1913) are in WO 97. To an extent, the arrival of the online electronic catalogue means that this no longer matters, but

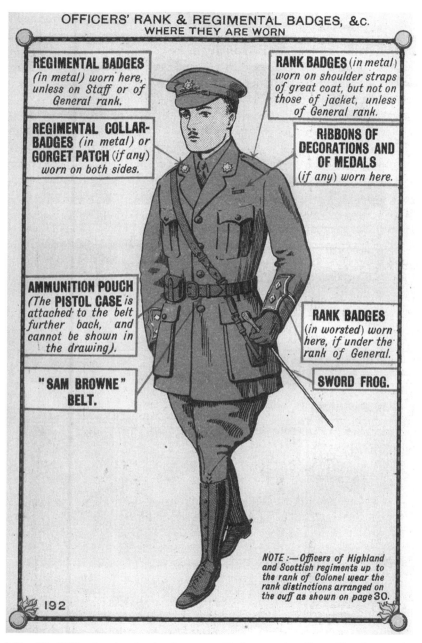

OFFICERS' RANK & REGIMENTAL BADGES, &c.
WHERE THEY ARE WORN

REGIMENTAL BADGES *(in metal) worn here, unless on Staff or of General rank.*

REGIMENTAL COLLAR-BADGES *(in metal) or* **GORGET PATCH** *(if any) worn on both sides.*

RANK BADGES *(in metal) worn on shoulder straps of great coat, but not on those of jacket, unless of General rank.*

RIBBONS OF DECORATIONS AND OF MEDALS *(if any) worn here.*

AMMUNITION POUCH *(The* **PISTOL CASE** *is attached to the belt further back, and cannot be shown in the drawing).*

RANK BADGES *(in worsted) worn here, if under the rank of General.*

"SAM BROWNE" BELT.

SWORD FROG.

NOTE :—Officers of Highland and Scottish regiments up to the rank of Colonel wear the rank distinctions arranged on the cuff as shown on page 30.

192

A First World War drawing showing the types of badges worn by officers. Identifying uniform badges can give a clue to a man's career.

occasionally it can be still useful to know how the records are organised, particularly if you need to use the manual finding aids when you visit Kew.

Many of the most important military records are now available only on microfilm. Where this is the case this is indicated in the appropriate place in the text and a list of the series of records which have been microfilmed is given in Appendix 4.

Microfilms are consulted in the Microfilm Reading Room, where they are all available in large cabinets for you to help yourself. The downside is that they are often difficult to read. It is also possible to print out copies for 25p per A3 sheet. Unfortunately in the experience of this researcher the quality of the printouts varies greatly. You may need to ask staff for help in printing out readable copies. The Microfilm Reading Room also has sets of the Army Lists and other reference guides on the shelves.

In the long term most popular records will be made available online, just as the census is at present. At the time of writing the First World War Medal Index Cards, Prerogative Court of Canterbury wills and a few other collections can be used at DocumentsOnline (www.nationalarchives.gov.uk/documentsonline). Other small series of records (mainly for the Great War) should be added shortly. Service records for First World War soldiers are likely to be added to a commercial website during 2007.

At Kew there is also an excellent bookshop selling a wide range of books on military and genealogical topics, a restaurant, computers where you can access the internet and a small museum.

1.5 Basic genealogical sources

It is easy to overlook the basic genealogical sources of birth, marriage and death records, census returns and wills in researching individual soldiers, but they are well worth checking out. And of course many researchers first become aware of having military ancestors from an entry in the census or on a marriage certificate.

Most of these records are now available online or likely to become so in the foreseeable future. Often the index can be searched for free, but you have to pay a small amount to download the information relating to an individual. Although this may seem expensive (particularly to a tightwad like the author) subscribing to one of these websites is probably cheaper than a journey to London or Edinburgh to see the originals for free. One downside is that the indexing, particularly to the censuses, is not always accurate, but with a bit of imagination and perseverance you should be able to track down your man.

Casualty returns and rolls created by the Army during the eighteenth and nineteenth centuries are discussed in Chapter 7, while records for the two world wars are described in Chapters 15 and 16.

1.5.1 Birth marriage and death records

National registration began in England and Wales on 1 July 1837 (Scotland – 1855, Ireland – 1864). The system has remained largely unchanged since then.

Apart from Scotland where the system is different, you need to order a certificate by using indexes and noting down a reference number which you quote when ordering a certificate. At the time of writing certificates cost £7 each (€6.98 in the Republic of Ireland, £10 Northern Ireland). Although the information contained naturally varies depending on the event, they all have occupations columns for new fathers and grooms, and the deceased should reveal whether they were soldiers or veterans.

There are several websites with indexes to English and Welsh certificates – the best known site is www.1837online.com, an alternative is www.familyrelatives. com. There is a free, if incomplete, alternative at www.freebmd.org.uk which is most useful for the period between 1837 and 1910. A complete set of registers is available at the Family Records Centre in London (address at end of chapter) if you don't want to use the online services. You can also order certificates there as well.

With the exception of southern Ireland it is now possible to order certificates online, as well as in person at their search rooms or by telephone. Visit the various General Register Office (GRO) websites for more information.

The Scots do things differently. In Scotland there are no indexes and the certificates over fifty years (seventy-five years marriage, 100 years births) are available online at www.scolandspeople.gov.uk A complete set (including certificates which are closed online) can be consulted at New Register House in Edinburgh. Fees are payable in both cases, but the good news is that they are more informative than their equivalents elsewhere in the British Isles.

Before 1837 (and indeed they are still kept today) baptisms, marriages and burials are recorded in a somewhat haphazard way by local clergymen in parish registers. Until 1812 men were not required to give occupations, although occasionally earlier registers may indicate whether a man was a soldier. The International Genealogical Index (IGI) is an incomplete nationwide index to births and marriages. It can be searched online for free at www.familysearch.org, and many archives and libraries have copies on microfiche. The equivalent for burials is the National Burial Index – some entries can be found at www.familyhistoryonline.co.uk, and libraries and archives may have copies on CD. Both indexes end at about 1837 with national registration.

1.5.2 Birth, marriage and death records: unique sources for the Army

General register offices. Each of the General Register Offices for the separate parts of the British Isles has small collections of birth, marriage and death records for soldiers. Of these the GRO in London was the most important as all information was gathered here to be passed on to Scottish and Irish offices as appropriate. If you can't find Irish or Scottish soldiers in the appropriate place in Dublin or Edinburgh then it is worth check here.

On the ground floor at the Family Records Centre (FRC) are the Chaplains' Returns between 1796 and 1880, which record births, baptisms, marriages and deaths of Army personnel and their families. They are generally for Army stations abroad. There is a similar series of registers between 1761 and 1924 for births, baptisms, marriages, deaths and burials of soldiers and their families at home and abroad.

The FRC also holds the Army Register Books which were kept from 1881 to 1959 containing births, marriages and deaths outside the UK. There is a combined register after 1959 for the three services. There are also indexes to deaths in the two world wars and the Boer War. They can be useful if you do not know the regimental number or unit a man served with, because these details are given in the registers. You can search the registers online at www.1837online.com or buy the indexes on CD from S&N www.sandn.net.

You may wish to buy certificates for deceased relations, but they are of limited use because they do not give cause or place of death.

An extract from the Army Chaplains' Returns at the Family Records Centre.

The General Register Office for Scotland has in its 'minor records' series Army Returns of births, deaths and marriages of Scottish persons at military stations abroad between 1881 and 1959. Later records are in the Service Departments' Registers. There are also certified copies of entries relating to marriages solemnised by Army chaplains outside the United Kingdom since 1892, where one of the parties to the marriage is described as Scottish.

Registers are also held for the South African War (1899–1902), which record the deaths of Scottish soldiers; the First World War – deaths of Scottish persons serving as non-commissioned officers or ordinary soldiers (but not officers) – and the Second World War, which comprises incomplete returns of the deaths of Scottish members of the Armed Forces. These records are due to be added to the Scotland's People website shortly, and of course can be seen at New Register House.

The General Register Office of Ireland in Dublin has similar records for Irish soldiers up until 21 December 1921. Later records for soldiers from the Six Counties can be found at the Northern Irish GRO in Belfast.

The National Archives. The National Archives has a small number of regimental registers of births, baptisms, marriages and burials for Militia units, in series WO 68. The exceptions are marriages and baptisms for Royal Artillery: (1817–1827, 1860–1877) in WO 69/551-582 and the Royal Horse Artillery (1859–1877) in WO 69/63-73.

Series WO 156 contains registers of births at Dover Castle (1865–1916, 1929–1940); Shorncliffe and Hythe (1878–1939); Buttervant (1917–1922); and Fermoy (1920–1921), burial registers for the Canterbury garrison (1808–1811, 1859–1884, 1957–1958) and baptisms and banns of marriage for Army personnel in Palestine between 1939 and 1947.

Registers of baptisms between 1691 and 1812, marriages (1691–1765) and burials (1692–1856) for the Royal Hospital Chelsea are in RG 4/4330-4331, 4387.

During the eighteenth and early part of the nineteenth centuries applicants for government jobs, including the Army, had to supply a certificate showing their place and date of baptism in order to prove their adherence to the Church of England. A collection of certificates for officers between 1777 and 1892 is in WO 32/8903-8920. Each piece is indexed. Series WO 42 contains certificates of birth, baptism, marriage and death, with wills, administrations, statements of services and personal papers of officers and their families, in cases where the officer died on service or on half pay and his widow applied for a pension, or application was made for a child's compassionate allowance, between 1755 and 1908 (although mostly between 1776 and 1881). There is a name index at the beginning of the class list.

Notifications to the War Office of marriages by officers between 1799 and 1882 are in WO 25/3239-3245.

1.5.3 Census returns

The first census was held in 1801 (Ireland 1821), but the first one which recorded details of individuals was not until 1841. More informative, however, are those between 1851 and 1901. The 1911 census not will be released until January 2012, although the one for Ireland is already open.

Census records tell you who was living at a particular address on census night with their full name and details of their relationship to the head of household, age, occupation and place of birth. Returns were made for barracks in the UK, but not for men overseas or on board troopships. They can be useful for tracking down soldiers' families or perhaps finding out a little more about the colleagues a man served with.

All the censuses are now available online at www.ancestry.co.uk. The 1881 census can searched for free at www.familyhistoryonline.co.uk, which also has a number of indexes to other censuses. In addition, there are several other pay-to-view sites. The 1841 census can be found at www.britishorigins.com, the 1861 census at www.1837online.com and the 1901 census is available at www.1901census.nationalarchives.gov.uk.

A national set of census records can be checked at the Family Records Centre and they also have free access to the online indexes.

Scottish census records are online at www.scotlandspeople.gov.uk and can be seen at New Register House in Edinburgh.

With the exception of 1901 and 1911, Irish census records have largely been destroyed; what survives can be seen at the National Archives of Ireland in Dublin. Apart from a few small indexes there is nothing yet online, although both the twentieth century censuses should be available by the end of 2008.

British troops overseas will appear in the appropriate colonial census, should one have been taken or survived. The Canadian census for 1881 is online at www.familysearch.org. The original records of the New South Wales census of 1828 are at TNA in HO 10/21-28, but have been published on CD.

1.5.4 Wills

It was natural for soldiers to make wills before going into action. Indeed during the two world wars the Army pay book, which was issued to all soldiers, included a simple will form which could be completed. In practice few other ranks made wills because in general they had very little personal property to bequeath to others except the immediate next of kin.

Before 1858 wills were proved in a bewildering variety of ecclesiastical courts; the most important of which was the Prerogative Court of Canterbury (PCC) in London. PCC wills are all indexed and available online at www.documentsonline. nationalarchives.gov.uk, so they are easy to check. Soldiers of all ranks had their wills proved here particularly during the eighteenth and early nineteenth centuries, such as James Jones, 'late Soldier in the First Regiment of Guards' whose will was proved in November 1770. At time of writing it costs £3.50 to download a will. The records can be seen for free at The National Archives and the Family Records Centre.

In 1858 a national system was set up with a network of district probate registries feeding in wills to the Principle Probate Registry in London. Calendars (or registers) are provided for each year. Copies on microfiche, between 1858 and 1946, are available in the Microfilm Reading Room at The National Archives and at other large archives or libraries. Copies of wills themselves cost £5 and can be ordered by post from the sub-registry in York.

Unfortunately you can't (yet) order or read post-1858 wills online.

It is understood that soldiers' wills for the two world wars are still with the Principal Probate Division, although they are not available for public inspection. It is hoped that they may eventually be transferred to The National Archives.

1.5.5 Wills: unique sources
At The National Archives, wills for many soldiers are in the casualty returns in WO 25, with a few wills for officers in WO 42. A related source is the Soldiers' Effects Ledgers between 1862 and 1880/1881 in WO 25/3475-3501.

Later ledgers, between April 1901 and March 1960, are with the National Army Museum. At present they are only open to public access for the period 1 April 1901–31 July 1914, although those for the First World War can be searched for a fee. The registers provide the following information: full name, regimental number and rank, date and place of death, place of birth, date of enlistment, trade on enlistment, next of kin as stated by regiment or on the individual's will, name of legatee, to whom authorised, and amount authorised.

The National Archives of Scotland has a collection of 30,000 wills for Scots soldiers covering the period between 1857 and 1966: the vast majority of which are for the First World War. There is an online catalogue, although you can only see digital images of the originals in the reading rooms. More details can be found at www.nas.gov.uk/guides/searchSoldiersWills.asp.

1.6 Private papers
Many officers and soldiers left accounts of their experiences in the Army in the forms of letters, diaries, photographs and autobiographies, which can really bring alive a campaign or battle in which your ancestor took part. It can be

difficult to track them down – even if they survive – and of course many remain with the families themselves.

Some accounts have been published and are well known, such as Frank Richards' *Soldiers Sahib* and *Old Soldiers Never Die* telling of his experiences as a private in the Royal Welch Regiment during the 1890s and 1900 and then in the First World War, but many others have either never been published or lie forgotten.

If an individual's papers have been deposited with an archive it should be possible to find where they are on the National Register of Archives at The National Archives. Fortunately the register is online (www.nationalarchives. gov.uk/nra) so it is easy to check.

The largest collection of private papers for the wars of the twentieth century is held by the Imperial War Museum. Catalogues for these collections are online at www.nationalarchives.gov.uk/a2a or less usefully at www.iwmcollections. org.uk.

The National Army Museum also has extensive collections which extend back into the eighteenth century: a list of their collections of private papers can be found at www.nationalarchives.gov.uk/nra.

Regimental museums collect papers of former members and may have published extracts in regimental journals and magazines, such as Margaret Kirwan's account of her time in the Crimea described on page 107.

1.7 Printed sources

There are a number of reference books, often available on the shelves of reference libraries, which contain something about an ancestor, particularly if he was a senior officer or had won a medal for gallantry.

Perhaps the most important biographical source is the *Oxford Dictionary of National Biography*, which most reference libraries will have online or in book form. Many local library services also allow holders of library cards to search this (and other sources) online for free, so it is worth asking whether yours does. There are nearly 3,000 entries for Army officers and other ranks (including four women soldiers who served disguised as men) and, of course, many other people whose biographies are included here also spent part of their career in the Army.

For the twentieth century it is worth checking entries in *Who was Who* which is published every ten years or so containing entries from the annual *Who's Who* for people who died in the previous decade. There is a cumulative index, which helps when you don't exactly know when somebody died. *Who's Who* first began publication in 1897 so is a great source for the twentieth century.

Many reference libraries have copies of the *Army List*, which lists all officers. Usually it will be a current edition, but they may have older lists stored behind the scenes. The staff should be able to tell you what is available.

1.8 Newspapers, magazines and journals

Newspapers are an underused resource for family history in general and for people researching soldiers, which is a pity because they can provide a wealth of information relating to:

- Individual soldiers and their service with the colours, particularly for men who won medals or who died while on active service. Officers generally receive better coverage than other ranks.
- Particular battalions and regiments, such as reports of regimental dinners or processions.
- Specific battles and campaigns. This may include letters from men who were there. They were often heavily censored (particularly from 1915 onwards) or rewritten for home consumption.

The best sources of information on individual soldiers, regiments and battalions can usually be found in local newspapers. The information is fullest for the Boer War and the two world wars, although you may well find obituaries for veterans of the Napoleonic and Crimean wars who died at a great age.

National newspapers do not generally contain as much about individual servicemen, although again officers tend to fare better. Soldiers are likely to be mentioned either if they won the Victoria Cross or other gallantry medals or had achieved anything unusual such as dying at an advanced age. General casualty lists were published (as well as lists relating to the award of medals and decorations for bravery to individuals), although they contain very few details beyond name, rank, and number.

The most accessible national newspaper is *The Times* which is fully indexed from its first issue in 1785. Indexes can be found in most reference libraries either on CD or online in The Times Digital Archive, which not only provides a fully searchable index but also digital images of each article. This is a superb resource.

The largest collection of newspapers and magazines is at the British Newspaper Library at Colindale in North London. There is an online catalogue, at www.bl.uk/catalogues/newspapers/welcome.asp, which will tell you what they have and the period it covers.

Most local studies libraries and record offices will have copies of newspapers for their areas on microfilm. The Newsplan project has provided local libraries with sets of local newspapers and where necessary microfilm readers on which to read them. More details can be found at www.newsplan2000.org and there are links to several regional websites listing which papers can be found where.

There are a number of projects to digitise newspapers and put them online, although it will be a few years before they are complete, and even then only a small proportion will ever be available.

Otherwise most local newspapers and where they are to be found are listed in Jeremy Gibson (et al), *Local Newspapers 1759–1920: A Select Location List* (2nd edition, Federation of Family History Societies, 2002).

The biggest problem is that newspapers have rarely been indexed, so you need to know a date when an event might have been reported. An obituary for a fallen soldier may, however, appear weeks or even months after his death, so you may need to go through many issues before you find what you are looking for.

Magazines are another important, if very neglected source. By the end of the nineteenth century most large companies had a staff magazine. There were also trade journals, magazines for enthusiasts and parish magazines. They often reported deaths and included letters from colleagues or parishioners on active service during the two world wars. Again the largest collection of these is at the British Newspaper Library, although parish magazines (where they survive) are likely to be at local studies libraries and record offices. The National Archives has sets of railway staff magazines.

By the First World War most regiments, or regimental associations, published a journal generally three or four times a year. Although the contents vary, they are likely to contain obituaries of old soldiers (particularly officers and senior NCOs), details of officers and men leaving the service, reminiscences and articles about life in the Army. It can be hard to track down copies, but regimental museums should have sets for their regiments and the Imperial War and National Army museums may also have copies.

There were a number of generally short lived weekly or monthly journals published for the Army, which contained details of promotions and postings as well as news stories and other material likely to appeal particularly to officers. They too can be hard to find, although the British Newspaper Library and the National Army Museum are worth approaching. The most long-lasting of these was the *United Service Journal and Naval Magazine* (*Colburn's United Service Magazine* from 1843) first published in 1829 and which lasted until 1890.

The National Archives has the occasional example, such as *June's Woolwich Journal* which was a newspaper for the Royal Artillery. It contained news of officers and their movements with other interesting information. Copies between 1847 and 1850 can be found in WO 62/48.

1.9 Parliamentary papers
Parliamentary papers were (and indeed are) published on behalf of Parliament. They are one of the sources least used by genealogists, which is surprising because they are well indexed and easy to use.

Of particular interest are reports by select committees (which often include verbatim interviews with individual witnesses, evidence etc.). These papers are on a variety of subjects from venereal disease among soldiers to punishment of

The Firepower Museum in Woolwich is devoted to the history of the Royal Artillery and has an excellent archive with lots of information about gunner ancestors.

other ranks and the sale of commissions by officers. Also of use are the various statistical returns made to Parliament by the War Office or other departments, for example there are a series of returns on the numbers of men who enlisted and then subsequently deserted.

Occasionally there are lists of particular officers and men. For example in 1739 and 1740, there are lists of officers on half pay as well as lists of officers on the Irish establishment (that is the Army in Ireland).

The National Archives has a complete set on microfilm. Large university and reference libraries should also have sets. A few libraries may also have them online on a subscription basis. There is an online index at www.bopcris.ac.uk and there are a variety of published indexes in book form or on CD.

1.10 Maps and photographs
Maps were (and are) an essential part of military planning. Hundreds of thousands of maps and plans have been drawn to show the lie of the land, how battles were fought and their results. The largest collection of such maps can be found at The National Archives. The major series is WO 78 – the first maps here date from 1627, and the series covers the world. To find them you can search TNA Online Catalogue or use the card index in the Map and Large Document Reading Room. Descriptions of some maps also appear in the series of published catalogues to the Archives' map holdings.

The National Archives has several small collections of maps for the First World War, although the best holdings of trench maps are at the Imperial

War Museum (IWM). You might want to look at Peter Chasseaud, *Topography of Armageddon: a British trench map atlas of the Western Front, 1914–1918* (Mapbooks, 1991), which shows the front line as it was in December 1917, to get your bearings before looking at more detailed trench maps.

The major series of maps for the Second World War is to be found in WO 252, along with reports about strategic targets. Otherwise maps may be found in war diaries or in intelligence or other files. Again the IWM also have a large collection.

The British Library's Map Library and the National Army Museum also have good collections of military maps. Regimental museums may also have smaller collections.

The largest collection of photographs and cinema film, particularly for the two world wars, is at the Imperial War Museum. At the heart of the museum's collections are the official photographs taken by British and Commonwealth photographers supplemented by material donated by former servicemen. The museum's Photographic Archive cannot tell you whether they have any pictures of your ancestor. However they welcome visitors searching the photographic collections although advance booking is necessary. A selection of images has been put online as part of the IWM's catalogue (www.iwmcollections.org.uk).

The National Army Museum and regimental museums have collections of photographs. The most popular subjects seem to be groups of officers, platoons of soldiers, or pictures of peacetime activities.

Useful addresses

The National Archives
Ruskin Avenue, Kew, Richmond, TW9 4DU
Tel: 020-8392 5200, www.nationalarchives.gov.uk

British Newspaper Library (Often called British Library Newspaper Library)
Colindale, London, NW9 5HE
Tel: 020-7412 7353, www.bl.uk/collections/newspapers.html
* The FRC is due to close by the end of 2008.

Family Records Centre
1 Myddelton St, London, EC1 1UW
Tel: 020-392 5300, www.familyrecords.gov.uk/frc

General Register Office
Postal address PO Box 2, Southport PR8 2HH
Search room: Family Records Centre, 1 Myddelton St, London, EC1 1UW
Tel: 0845 603 7788, email: certificate.services@ons.gov.uk,
www.gro.gov.uk, www.familyrecords.gov.uk/gro.

General Register Office of Ireland
Postal address: Civil Registration Office, Office of the Registrar General, Government Offices, Convent Road, Roscommon, Co. Roscommon.
Search room: Joyce House, 8/11 Lombard Street East, Dublin, 2.
Tel: +353 (0)90 663 2900, www.groireland.ie.

General Register Office (Northern Ireland)
Oxford House, 49–55 Chichester Street, Belfast, BT1 4HL
Tel: 028-90 252000, email: gro.nisra@dfpni.gov.uk, www.groni.gov.uk.

General Register Office for Scotland
New Register House, Edinburgh, EH1 3YT
Tel: 0131-334 0380, email: records@gro-scotland.gov.uk,
www.gro-scotland.gov.uk; www.scotlandspeople.gov.uk.

Imperial War Museum
Lambeth Road, London, SE1 6HZ
Tel: 020-7416 5000, www.iwm.org.uk.

National Archives of Ireland
Bishop Street, Dublin 8, Ireland
+353 (0)1 407 2300, email mail@nationalarchives.ie, www.nationalarchives.ie.

National Archives of Scotland
HM General Register House, 2 Princes Street, Edinburgh, EH1 3YY
Tel: 0131-535 1314, www.nas.gov.uk.

National Army Museum
Royal Hospital Road, London, SW3 4HT
Tel: 020-7730 0717, www.national-army-museum.ac.uk.

Probate Service (Principal Probate Registry)
Postal address: York Probate Sub-Registry, Duncombe Place, York, YO1 2EA
Search room: First Avenue House, 44–49 High Holborn, London, WC1V 6NP
Tel: 020-7947 6000, www.hmcourts-service.gov.uk/cms/wills.htm.

Further reading
There are dozens of introductory guides to family history. They all cover much the same ground, but these are the ones recommended by the author.

Anthony Adolph, *Collins guide to family history* (2nd edition, Collins, 2005)
Dave Annal, *Easy Family History* (The National Archives, 2005)
Simon Fowler, *The Joys of Family History* (Public Record Office, 2001)

Chapter 2

ORGANISATION OF THE ARMY IN THE EIGHTEENTH AND NINETEENTH CENTURIES

Until the First World War, the main fighting arms were foot soldiers, collectively called the infantry, and horse soldiers, collectively called the cavalry. These were supported by artillery, whose personnel served and fired guns of various calibres, and engineers, who shaped and supplied the battlefield by building and measuring and sapping and mining. Behind the scenes supporting the fighting men were the Commissariat, the Army Service Corps and the medical services.

2.1 Regiments
The British Army, as we know it today, effectively was created after Charles II's return from exile in 1660. A small number of regiments were established on a permanent basis to protect the king, based on units which had already been in existence during or before the Civil War. Their origins and the debate over which is the oldest regiment in the British Army is discussed at www.regiments.org/about/faq/oldest.htm.

It was at this time that the distinction between 'regiment' and 'battalion' became fixed. The battalion became the effective infantry fighting unit, and the regiment became an administrative organisation. However the terms have remained interchangeable, which has led to the occasional confusion.

For three hundred years the heart of the British Army has been the regiment. For many officers and men it has been a surrogate family which provided everything from comradeship and a focus for loyalty to a square meal each day.

The historian G M Trevelyan observed in his *Illustrated English Social History*, published in 1949, that the relationship was rural in origin:

> *The regiment was a society made up of grades answering to the social demarcations of the English village whence men and officers had come. It has been observed that when an ensign fresh from Eton was handed over to the respectful care and tuition of the colour sergeant, the relation of the two closely resembled that to which the younger man had been accustomed at home, when the old gamekeeper took him out afield to teach the management of his fowling piece and the arts of approaching game.*

Although it is possible to take this analogy too far, it is clear that many officers regarded war, at least until 1914, as being little more than an extension of hunting and shooting. And agricultural labourers were the preferred social class from whom recruits were chosen, their stolid nature being much preferred over flightier and mouthier town birds.

No two regiments were the same. Each one jealously maintained its idiosyncrasies which, although occasionally annoying to the authorities, had usually a loyalty and a fighting spirit second to none. It helped that the regiment recruited from a particular area and that a surprising number of sons followed their fathers and grandfathers into the unit. Every Duke of Wellington (apart from the 7th and current Duke), since the first 'Iron Duke', for example, has served as an Army officer.

New recruits were rigorously taught the traditions and history of the regiment. Even today after decades of reorganisation and amalgamation, the sense of individuality remains. There were huge protests in Scotland in 2004 when it was proposed to merge all the Scottish regiments into one 'super' Royal Regiment of Scotland.

Although undoubtedly proud of their traditions and history, inevitably the same sense of individuality is missing from the Royal Artillery, Royal Engineers and the other branches of the Army.

From the point of view of the family historian, the regiment is important because many records were either kept, or arranged, by regiment. It was not until 1920 that an Army-wide service number was introduced. Before then records were maintained by a regimental record office. For more about this see Appendix 1. Even today many records, which for other units are at The National Archives, are still maintained by the Household Cavalry.

Infantry regiments were formed as and when required, and were originally denominated by the name of the colonel (the most senior regimental officer, and usually the man who raised and equipped the regiment), which meant that names of units changed as colonels came and went.

Army uniforms in the 1890s showing the wide range of dress still worn by regiments.

To avoid confusion regiments began to be numbered sequentially, 1 Regiment of Foot, 2 Regiment of Foot and so on. In this way it was possible to refer to the regiments as either the 1 Regiment, or the 1 Foot.

Many regiments were also given names, thus the 1 Foot was called the Royal Scots, the 2 Foot the Queen's Regiment, and the 3 Foot The Buffs. The latter is an interesting example of how something which looks odd today arose from a commonsense solution to a problem. The 3 and 19 Regiments of Foot had, in the eighteenth century, colonels called Howard. The 3 Foot had buff facings (the colour of the reveal on collars and cuffs of the red coat worn by the infantry at this time), and the 19 Foot had green facings, so to distinguish them, unofficially the 3rd became the Buff Howards and the 19th the Green Howards.

In order to aid recruitment, it was decided in 1782 that regiments of foot should be given territorial designations. The 28 Foot for example became 28 (North Gloucestershire) and the 67 Foot became 67 (South Hampshire). Even today the Cheshire Regiment is still formally known as the 22 (Cheshire) Regiment. This association was at best tenuous, particularly as regiments recruited as, and when and where, they could, especially in Ireland.

Almost exactly 100 years later in 1881, a more successful attempt was made to link regiments to locality. Now, each regiment was re-titled with a territorial (usually county) name, based at a local regimental depot (for the smaller counties this would normally be the county town), and integrated managerially with the local militia and volunteer units.

In Yorkshire, for example, the 1st and 2nd Battalions of the 19 Regiment of Foot, renamed Alexandra, Princess of Wales's Own (Yorkshire Regiment) were based at Richmond. In Cornwall, the old 32 (Cornwall) Light Infantry and the 46 (South Devonshire) Foot became the 1st and 2nd Battalions, The Duke of Cornwall's Light infantry based at Bodmin. A list of the merged regiments and their titles is given at the end of the chapter.

The Foot Guards were regarded as the elite infantry units, and had high status; their original function was to guard the sovereign, although they fought as normal line (or heavy) infantry when on campaign. Officers of the Guards regiments held double rank: that is a captain in the Guards ranked as a major elsewhere in the Army. Ordinary guardsmen were always taller and generally fitter than their equivalents in regiments of the line.

The Line Regiments comprised the bulk of the infantry, and were so called because the regiments paraded in line in order of seniority from right to left. These regiments formed the backbone of the Army when it saw action.

The Light Infantry battalions, such as the 51st Light Infantry, originally were used for scouting, flanking and outpost duties, and had slightly different equipment. After the Napoleonic Wars only the name remained (Durham Light Infantry, Somerset Light Infantry etc.) as they became ordinary infantry regiments, although it is interesting to note that the first British troops to land on D-Day came from the Ox and Bucks Light Infantry, undertaking a role that would have been familiar to their predecessors under Wellington.

The Rifle Brigade (the 95 Foot) had a specialised function as sharpshooters using rifled weapons as opposed to smooth-bore muskets, and were used as skirmishers, wearing green uniforms with black equipment for camouflage purposes, hence their nickname of the Green Jackets which is still in use today.

Regiments and battalions have been formed, merged and disbanded many times over the years. And to make it worse names have constantly changed, which can make things confusing for the amateur researcher. However, help is at hand in the form of Arthur Swinson's *A Register of the Regiments and Corps of the British Army: the Ancestry of the Regiments and Corps of the Regular Establishment* (Archive Press, 1972). Slightly more up to date is Christopher Chant, *A Handbook of British Regiments* (Routledge and Kegan Paul, 1988). Reference libraries should have one or both of these volumes. The best online source is www.regiments.org, although it is incomplete.

Infantry battalions (abbreviated as Bn or Bns) were usually divided into ten Companies (abbreviated to Coy or Coys). In the Napoleonic wars, each battalion had one Grenadier company, a Light company and eight Battalion companies. The Grenadier and Light companies from different battalions could be amalgamated for storming, or scouting and outpost duties respectively, as required.

At full strength a battalion numbered 1,000 men (today it is about 500). A battalion was broken down into companies which could number 100 men each, normally under the command of a captain. After 1913, companies in turn were subdivided into platoons, consisting of between seven to ten men, generally under a lieutenant or increasingly a non-commissioned officer such as a sergeant or corporal. If you are interested, simple definitions of the composition and origins of British Army units can be found at www.regiments.org/regiments/nomencla.htm.

However, the number of men actually in each battalion could vary considerably. For example, at Waterloo, the 2/73 Foot (that is, the Second Battalion of the 73 Regiment of Foot) comprised 558 other ranks, whereas the 2/3 Guards numbered over 1,061 and the 2/44th mustered as few as 455.

Numbers varied for many reasons: loss in battle, sickness (particularly common in the tropics), desertion, men detached on other duties, such as recruiting parties, and the failure to recruit at home. As a fighting unit a battalion could be decimated in particularly heavy fighting, as occurred during the First World War. The 1st Battalion, Hampshire Regiment, for example, lost twenty-six officers and 559 soldiers during the opening day of the Somme (1 July 1916). Its war diary for the day laconically contains the following entry: 'The casualties on officers amounted to 100% and was also heavy in the other ranks.' (TNA reference WO 95/1495.)

Before 1878, the muster rolls in series WO 12 at Kew tell you month by month the strength of a particular battalion. Battalions were brought together into brigades (two or more battalions) and brigades were brought together in divisions. Infantry divisions usually included their own support services, cavalry divisions included theirs, and these were all described in Orders of Battle, which recorded the organisation of the Army at any given time. Monthly Army Returns can be found at Kew in series WO 73, between 1859 and 1960. Earlier records are in WO 17 (1754–1866).

2.2 Cavalry

Like the infantry, the cavalry consisted broadly of 'heavy' and 'light' regiments, divided into squadrons which comprised two troops, but cavalry regiments comprised no more than 500 or so soldiers, sometimes many fewer. Heavy cavalry included both the Household Cavalry (Horse Guards and Life Guards), and regiments of horse called Dragoons and Dragoon Guards. Dragoons were originally mounted infantry, using horses as a mode of transport to the fighting area, where they dismounted and fought on foot. By the time of the Napoleonic Wars this definition had ceased to have any relevance.

However, the difference between Dragoon Guard and Dragoon regiments was very important, and for family historians the failure to distinguish between, say, the Fourth Dragoons and the Fourth Dragoon Guards would mean researching the wrong regiment entirely.

Cavalry regiments, however, were not geographically centred. Many cavalry regiments were disbanded or merged in 1922. This year also marked the beginning of their transfer into armoured units, the tank replacing the horse, except on rare ceremonial occasions. The last cavalry charge in the British Army actually took place during the Battle of El Mughar in Palestine on 13 November 1917. The Buckinghamshire Hussars, supported by the Dorset and Berkshire Yeomanry Regiments, overran a Turkish position, taking several hundred prisoners.

2.3 Other branches

Artillery was divided into three main categories: field artillery, siege artillery and garrison artillery, with the former being divided into horse and foot (essentially horse artillery supported cavalry formations and foot artillery supported infantry formations). The artillery was organised on a brigade (troop in the Horse Artillery) and company basis, the companies taking their name from their commanding officer. The Royal Artillery was relatively small compared with the artillery of continental armies.

Until 1855 the Corps of Royal Engineers was managed by the Board of Ordnance, and was technically not part of the Army at all. The Corps consisted entirely of officers at this period, and was very small numerically. The physical work was originally carried out by the Royal Military Artificers, but there were so few of them that in 1812 the whole establishment was reorganised and renamed the Royal Sappers and Miners, and divided into companies commanded by Royal Engineer officers. The two corps finally merged in 1856.

2.4 Campaign records before 1914

Until 1908 there was no permanent general staff. Planning for campaigns was done ad hoc, which helps to explain why so many wars (especially the Crimean War) were badly organised and managed. This also affects records of the campaigns themselves, which you should consult if, for example, you are interested in a particular action in which an ancestor fought.

Until the mid-eighteenth century the most useful sources can be found in SP 8, SP 41, SP 44, SP 87 and, for actions in America and the West Indies, CO 5 and CO 318. Many State Papers have been published in the Calendars of State Papers Domestic Series, Charles II, James II, William and Mary, Anne (Public Record Office Calendars, London 1860–1947). These calendars are well indexed.

For a century from about 1732 to the 1840s material can also be found in WO 1, WO 28, WO 36 and WO 40. The best place to look for material for campaigns after 1853 is WO 33, with some material in WO 32.

Maps and plans for campaigns from the seventeenth century onwards can be found in WO 78.

Many official papers ended up in the private papers of senior officers and if you plan to do a detailed study you will need to consult them. For example the original of Wellington's despatch announcing the great victory at Waterloo is in his papers at the British Library. The National Archives has papers of a few commanders who participated in the American War of Independence in PRO 30/55. The National Register of Archives, at Kew (or online at www.nationalarchives.gov.uk/nra.) has lists of where most collections of papers are to be found.

Before you start researching a particular battle or campaign it is a good idea to read up on the subject. Books (and to a much lesser extent the vast majority of websites) will put the event in which you are interested in context and if it is a half-way decent publication there will be footnotes and a bibliography which allow you to go to the original sources consulted by the author. A model of this kind is Thomas Pakenham's *The Boer War* (Weidenfeld and Nicolson, 1979) which combines all this and is very readable as well.

Although often ignored by historians, commanding officers prepared detailed accounts of campaigns, known as despatches, for the War Office and later published in the *London Gazette* and often in newspapers, such as *The Times*, as well. The names of officers and men who made a significant contribution to the war or campaign were listed. These 'mentions in despatches' became, in effect, the lowest level of medal, although it was not until 1920 that a formal award was made.

Until well into the nineteenth century records relating to campaigns can be found in a variety of different places at The National Archives (and elsewhere). It is well worth typing the name of an action or battle, such as Waterloo or Rorke's Drift, into TNA's online catalogue to see what turns up.

TNA also provide two useful online research guides *British Army: Campaign Records, 1660–1714*, and *British Army: Campaign Records, 1714–1815*. A general guide to the War Office and its workings is Michael Roper, *Records of the War Office and Related Departments, 1660–1964* (PRO, 1998).

The best published guide to sources at The National Archives, however, remains the *Alphabetical Guide to War Office and other Military Records* (Public Record Office Lists and Indexes vol. LIII, 1931; Kraus Reprints, 1963). This is an index by subject, name and regiment to the appropriate State Papers (before about 1782) and War Office records. It gives specific document references, as well as some very eccentric and amusing entries.

British Army Order of Precedence

Cavalry and Guards (in existence 1914)

Life Guards
Royal Horse Guards
1 (King's) Dragoon Guards
2 (Queen's Bays) Dragoon Guards
3 (Prince of Wales) Dragoon Guards
4 (Royal Irish) Dragoon Guards
5 (Princess Charlotte of Wales)
   Dragoon Guards
6 Dragoon Guards (Carabiniers)
7 (Princess Royal's) Dragoon Guards
1 (Royal) Dragoons
2 Dragoons (Royal Scots Greys)
3 (King's Own) Hussars
4 (Queen's Own) Dragoons
5 (Royal Irish) Lancers
6 (Inniskilling) Dragoons
7 (Queen's Own) Hussars
8 (King's Royal Irish) Hussars
9 (Queen's Royal) Lancers

10 (Prince of Wales' Own Royal)
   Hussars
11 (Prince Albert's) Hussars
12 (Prince of Wales Royal) Lancers
13 Hussars
14 (The King's) Hussars
15 (The King's) Hussars
16 (The Queen's) Lancers
17 (Duke of Cambridge's Own)
   Lancers
18 Hussars
19 Hussars
20 Hussars
21 Hussars
Grenadier Guards
Coldstream Guards
Scots Guards
Irish Guards
Welsh Guards

Infantry regiments (1881–1958)

Post 1881 title	Merged from
Royal Scots (Lothian Regiment)	
Queen's (Royal West Surrey Regiment)	
The Buffs (East Kent Regiment)	
King's Own (Royal Lancaster Regiment)	
Northumberland Fusiliers	
Royal Warwickshire Regiment	
Royal Fusiliers (City of London Regiment)	
King's (Liverpool Regiment)	
Norfolk Regiment	
Lincolnshire Regiment	

Post 1881 title	Merged from
Devonshire Regiment	
Suffolk Regiment	
Prince Albert's (Somerset Light Infantry)	
Prince of Wales's Own (West Yorkshire Regiment)	
East Yorkshire Regiment	
Bedfordshire Regiment	
Leicestershire Regiment	
Royal Irish Regiment	
Alexandra, Princess of Wales's Own (Yorkshire Regiment)	
Lancashire Fusiliers	
Royal Scots Fusiliers	
Cheshire Regiment	
Royal Welsh Fusiliers	
South Wales Borderers	
King's Own Borderers	
Cameronians (Scottish Rifles)	26 Cameronians/90 Perthshire Volunteers Light Infantry
Royal Inniskilling Fusiliers	27 Inniskilling Regiment/ 1st Btn (Madras Infantry) Regiment
Gloucestershire Regiment	28 North Gloucestershire Regiment/ 61 South Gloucestershire Regiment
Worcestershire Regiment	29 Worcestershire Regiment/ 36 Herefordshire Regiment
East Lancashire Regiment	30 Cambridgeshire Regiment/ 59 2nd Nottinghamshire Regiment
East Surrey Regiment	31 Huntingdonshire Regiment/ 70 Surrey Regiment
Duke of Cornwall's Light Infantry	32 Cornwall Regiment Light Infantry/ 46 South Devonshire Regiment
Duke of Wellington's (West Riding Regiment)	33 (Duke of Wellington's) Regiment/ 76 Foot
Border Regiment	34 Cumberland Regiment/ 55 Westmorland Regiment
Royal Sussex Regiment	35 (Royal Sussex) Regiment/ 107 Bengal Infantry

Post 1881 title	Merged from
Hampshire Regiment	37 North Hampshire Regiment/ 67 South Hampshire Regiment
South Staffordshire Regiment	38 1st Staffordshire Regiment/ 80 Staffordshire Volunteers Regiment
Dorsetshire Regiment	39 Dorsetshire Regiment/ 54 West Norfolk Regiment
Prince of Wales's Volunteers (South Lancashire Regiment)	40 2nd Somersetshire Regiment/ 82, Prince of Wales's Volunteers Regiment
Welsh Regiment	41 Welsh Regiment/ 69 South Lincolnshire Regiment
Black Watch (Royal Highlanders)	42 Royal Highland Regiment/ 73 Perthshire Regiment
Oxfordshire Light Infantry	43 Monmouthshire Light Infantry/ 52, Oxfordshire Light Infantry
Essex Regiment	44 East Essex Regiment/ 56 West Essex Regiment
Sherwood Foresters (Derbyshire Regiment)	45 Nottinghamshire (Sherwood Foresters) Regiment/ 95 Derbyshire Regiment
Loyal North Lancashire Regiment	47 Lancashire Regiment/81 Loyal Lincolnshire Volunteers Regiment
Northamptonshire Regiment	48 Northampton Regiment/ 58 Rutlandshire Regiment
Princess Charlotte of Wales's (Berkshire) Regiment	49 Hertfordshire (Princess Charlotte of Wales's) Regiment/ 66 Berkshire Regiment
Queen's Own (Royal West Kent Regiment)	50 Queen's Own Regiment/ 97 Earl of Ulster's Regiment
King's Own (Yorkshire Light Infantry)	51 (Yorkshire West Riding) King's Own Light Infantry/ 105 (Madras Light Infantry) Regiment
King's (Shropshire Light Infantry)	53 Shropshire Regiment/ 85 Bucks Volunteers – King's Light Infantry
Duke of Cambridge's Own (Middlesex Regiment)	57 West Middlesex Regiment/ 77 East Middlesex Regiment

Post 1881 title	Merged from
King's Royal Rifle Corps	60 King's Royal Rifle Corps
Duke of Edinburgh's (Wiltshire Regiment)	62 Wiltshire Duke of Cambridge's Own Regiment/ 99 (Duke of Edinburgh's) Regiment
Manchester Regiment	63 West Suffolk Regiment/ 96 Foot
Prince of Wales's (North Staffordshire Regiment)	64 2nd Staffordshire Regiment/ 98 (Prince of Wales's) Regiment
York and Lancaster Regiment	65 (North Riding) Regiment/ 84 York and Lancaster Regiment
Durham Light Infantry	68 Durham Light Infantry/ 106 Bombay Light Infantry
Highland Light Infantry	71 Highland Regiment Light Infantry/ 74 Highlanders
Seaforth Highlanders (Ross-shire Buffs, Duke of Albany's)	72 Duke of Albany's Own Highlanders/ 78 Ross-shire Buffs
Gordon Highlanders	75 Stirlingshire Regiment/ 92 Foot Gordon Highlanders
Queen's Own Cameron Highlanders	79 Foot/Queen's Own Cameron Highlanders
Royal Irish Rifles	83 (County of Dublin) Regiment/ 86 Royal County Down, Regiment
Princess Victoria's (Royal Irish Fusiliers)	87 Royal Irish Fusiliers/ 89 (Princess Victoria's) Regiment
Connaught Rangers	88 Foot Connaught Rangers/94 Foot
Princess Louise's (Argyll and Sutherland Highlanders)	91 (Princess Louise's Argyllshire) Highlanders/93 (Sutherland Highlanders)
Prince of Wales's Leinster (Royal Canadians)	100 Prince of Wales's Royal Regiment (Canadian Regiment)/ 109 (Bombay Infantry)
Royal Munster Fusiliers	101 Royal Bengal Fusiliers/ 104 Bengal Fusiliers
Royal Dublin Fusiliers	102 (Royal Madras) Fusiliers/ 103 (Royal Bombay) Fusiliers
Rifle Brigade (Prince Consort's Own)	Prince Consort's Own Rifle Brigade

Corps in existence in 1920

Machine Gun Corps
Queen Alexandra's Imperial
   Military Nursing Service
Royal Army Chaplains
Royal Army Medical Corps
Royal Army Ordnance Corps
Royal Army Pay Corps
Royal Army Service Corps
Royal Army Veterinary Corps

Royal Artillery (including Royal Field
   Artillery, Royal Garrison Artillery,
   Royal Horse Artillery)
Royal Corps of Signals
Royal Engineers
Royal Military Police (including corps
   of Military Mounted and Military
   Foot police)
Women's Auxiliary Army Corps

Chapter 3

THE ARMY BEFORE 1660

Researching ancestors who were in the Army before 1660, is difficult for two reasons. The first is that the Army did not exist as we know it today – units were raised as and when needed – and secondly there was no real system of record keeping until the sixteenth century at least, and no real documentation of individual service until the eighteenth century.

Initially military service was part of the feudal obligation paid to the lord of the manor; he who might be required to raise a certain number of men, based on the size of his estates, to serve the sovereign on campaign abroad. Records can be found in a number of series at The National Archives, including C 54, C 64 (including material about Agincourt), C 65 and E 101. The system survived until the mid-eighteenth century in Scottish clans, where crofters were expected to fight for the clan chief when required.

More information can be found in the TNA Research Guide *Medieval and Early Modern Soldiers*. The guide also includes an exhaustive reading list. A good general introduction to researching ancestors who lived during the period is Paul Chambers, *Medieval Genealogy* (Sutton, 2005). This is complemented by an excellent website www.medievalgenealogy.org.uk, which includes a number of pages and resources devoted to the military.

3.1 Militia records

From Anglo-Saxon times, men were also liable to military service as a local home defence force or Militia. The Statute of Winchester of 1285 required all those between fifteen and sixty to be assessed to equip themselves with weapons and armour, according to their means, from scythes and knives for those holding less than forty shillings worth of land to horse and armour for the wealthiest individuals.

In the sixteenth century, local Commissioners of Array, or the Lord Lieutenant of the county, assisted by local officials such as the parish constable, were responsible for making and inspecting such assessments. They can therefore give very valuable information for many economic, local and family historians, as well as for those interested in military history. L Boynton *The Elizabethan Militia, 1558–1638* (London, 1967) describes the system in more detail.

Muster rolls list the names of local inhabitants who were liable to military service and the equipment they were required to have. They are by no means complete. Historians estimate that about a third of men were excluded for one reason or another.

The earliest surviving one dates from 1522. Some rolls, or certificates of musters giving only total numbers of men, were forwarded to the Exchequer or the Privy Council. These records, now at TNA, can be found scattered through series E 36, E 101, E 315,

SP 1 and SP 2. Later ones for the reigns of Elizabeth I, James I and Charles I are in SP 12, SP 14, SP 16 and SP 17. More details are given in an online research guide *Tudor and Stuart Militia Rolls*.

Other records have only survived with the private papers of those local gentry families who served as commissioners of array or deputy lieutenants and these may be in local record offices.

The White Tower at the Tower of London, for centuries of home of English military power.

However, the private papers of a Cheshire gentleman, John Daniel of Daresbury, at The National Archives in SP 46 or SP 52, contain correspondence and papers relating to musters and commissions, and several muster rolls of the trained band of which he was captain.

To find which records are held where, you need to use Jeremy Gibson and Alan Dell's *Tudor and Stuart Muster Rolls – A Directory of Holdings in the British Isles* (Federation of Family History Societies, 1991). Arranged by county, and then by hundred, wapentake, lathe, etc. (that is a sub-division of a county), it gives:

- what is held by The National Archives, with full document references
- what is held by local record offices
- what has been transcribed and published

Many publications by local record societies are held by The National Archives Library. In addition a few lists have been transcribed and placed online.

After the Civil War, the Militia was in abeyance until 1757 (see Chapter 10).

3.2 The English Civil War

There are no individual service records. There are, however, numerous references to serving soldiers (generally officers) in the public records that may be identified given time, notably the SP, PC, AO 3 and E 101 series. The best place to begin a general search for an officer is with the printed Calendars of State Papers Domestic, particularly the *Calendar of the Committee for Compounding with Delinquents* (which levied fines on royalists) and the *Calendar of the Committee for the Advance of Money*. They all have name indexes.

Edward Peacock's *Army Lists of the Roundheads and Cavaliers* (London, 1863) lists officers who were serving in 1642 in the royalist and parliamentary armies. A copy is available at The National Archives.

Sources relating to royalist officers of the rank of major and above are described in 'The Royalist Officers Corps 1642–1660' by P R Newman, *Historical Journal*, 26 (1983). For commissions granted by the king to raise regiments and appoint senior officers, see W H Black (ed.) *Docquets of Letters Patent 1642–6* (1837). Commissions granted to junior officers by commanding officers are less easy to trace, although may sometimes be found in the private papers of local gentry families which may have been deposited in a local record office.

After the Restoration, a fund of £60,000, supplemented by a tax on office-holders, was set up, to be distributed among 'such truly loyal and indigent officers who have had real command of soldiers according to their several commissions and who have never deserted his Majesty during the late times of rebellion and usurpation, and have not a sufficient livelihood of their own'. Over

Sergeant Major Pollard.[76]
Captaine Bulhead.
Captaine Prowſe.
Captaine Thomas.

Captaine Colesfoote.
Captaine Atkinſon.
Captaine Bateman.
Captaine Denby.

11 REGIMENT.

Sir Lewis Dives, Colonell.[77]
Liev. Colonel Lucy.
Sergeant Major Withrington.[78]
Captaine Browne.

Captaine Thomas Furbuſh.
Captaine Ley.
Captaine Johnſon.
Captaine Slingſby.[79]

In " a liſt of the priſoners of Quality now ſecured in Cheſter," publiſhed in the *Mercurius Politicus*, Sept. 1-8, 1659, occur " Lord Kilmorey " and " Mr. Thomas Nedham, Brother to Lord Kilmorey."

[75] Sir Faithful Forteſcue was colonel of the third troop of horſe raiſed for the expedition into Ireland, 1642. He was with his troop draughted into the Parliamentary Army under the command of the Earl of Eſſex. At Edge Hill battle, Sir Faithful Forteſcue with his whole troop left the Parliamentary Army " and preſented himſelf and his troop to Prince Rupert .... The ſudden and unexpected revolt of Sir Faithful Forteſcue with a whole troop .... had not ſo good fortune as they deſerved ; for by the negligence of not throwing away their orange tawney ſcarfs, which they all wore as the Earl of Eſſex's colours, and being immediately engaged in the charge, many of them, not fewer than ſeventeen or eighteen, were ſuddenly killed by thoſe to whom they had joined themſelves."—CLARENDON'S *Rebellion*, pp. 308, 309.
Arms, azure, a bend engrailed, argent, cotized, or.

[76] Pollard, Sir Hugh, ſlain at Dartmouth, Jan. 18, 1646.

[77] Wounded at Worceſter, Sept. 23, 1642. Made priſoner at the taking of Sherbourne Caſtle, of which he was governor, Aug. 15, 1645. Eſcaped from cuſtody, Jan. 30, 1649.

[78] Sir William Widdrington, firſt Baronet. Created Baron Widdrington of Blankney, co. Lincoln, Nov. 10, 1643. " He was one of the firſt who raiſed both horſe and foot at his own charge and ſerved eminently with them under the Earl of Newcaſtle."—CLARENDON'S *Hiſt.* p. 763.
Killed at the battle of Wigan, Lancaſhire, Aug. 27, 1651.

[79] Sir Henry Slingſby, Bart. of Scriven, co. York. A member of the Parliament of 1640. Defeated by Sir Hugh Cholmley at Guiſborough, Jan. 16, 1643. Taken priſoner in Cornwall, Jan. 1650. Impriſoned in Pendennis and Exeter Caſtles. Tried by a Court of High Commiſſion. Beheaded on Tower

12 REGIMENT.

Colonell Sir Charles Lucas.[80]
Liev. Colonell Stanley.
Sergeant Major Kelley.
Captaine Hodges.

Captaine Ford.
Captaine Burley.
Captaine Strangewayes.
Captaine Whiteaway.

13 REGIMENT.

Colonell Sir George Gotherick.
Lievtenant Colonell Waſhington.
Sergeant Major Powell.
Captaine Iſaack.

Captaine Johnſon.
Captaine Lever.
Captaine Burrowes.
Captaine Sutton.

14 REGIMENT.

Colonell Oſborne.[81]
Liev. Colonell Savage.

Sergeant Major Oneale.[82]
Captaine Forſter.

Hill, June 8, 1658. An intereſting diary, written by Sir Henry Slingſby, has been preſerved and twice printed. The beſt edition is that edited by D. Parſons, M.A. 8vo. 1836.

[80] Sir Charles Lucas, elder brother (born of the ſame parents, but before wedlock) of John, firſt Baron Lucas of Shenfield, co. Eſſex, and next heir of his brother's barony and eſtates.—NICHOLAS'S *Hiſtoric Peerage*, 1857, p. 301.
Tried by Court Martial and ſhot after the ſiege of Colcheſter, Aug. 28, 1648. " He was very brave in his perſon, and in a day of battle a gallant man to look upon and follow ; but at all other times and places of a nature not to be lived with, of an ill underſtanding, of a rough and proud nature, which made him during the time of their being in Colcheſter more intollerable than the ſiege."—CLARENDON'S *Hiſt.* pp. 664, 5.

[81] Edward Oſborne of Kiveton, co. Notts, Knight. Created a baronet, July 13, 1620. Vice-Preſident of the North of England, 1629. Father of Thomas Oſborne, firſt Duke of Leeds.—THORESBY'S *Ducatus Leodienſis*, p. 2.

[82] Daniel O'Neill was active in oppoſition to the Earl of Strafford. Committed to the Tower by the Parliament, from whence he eſcaped in women's clothing, and fled to the Low Countries. He returned and joined the

D

A page from Edward Peacock's Army Lists of the Roundheads and Cavaliers *(London, 1863) lists officers who were serving in 1642.*

5,000 claimants were published in *A List of Officers Claiming to the Sixty Thousand Pounds*, a copy of which can be found in SP 29/68. It is indexed.

There are no comprehensive lists of officers in the parliamentary forces. R R Temple's 'The Original Officer List of the New Model Army', *Bulletin of the Institute of Historical Research* (59) 1986 publishes a list of officers in the New Model Army in March 1645, from an original in the House of Lords Parliamentary Archives Record Office. There are numerous references to individuals in C H Firth and G Davies, *The Regimental History of Cromwell's Army* (Oxford, 1940) and I Gentles, *The New Model Army* (Blackwell, 1992).

There are many warrants, accounts and muster rolls relating to the payment of parliamentary soldiers in SP 28. These records are not included in the published Calendars, nor are there any name indexes. SP 28/265 contains not only a muster of the fifty-five officers and men of Captain Giles Hicke's cavalry troop but a separate list of 'distressed widowes whose husbands were slaine in the service'. Payments made to local county forces may be recorded in the accounts of the county committees, also in SP 28. Some certified accounts of soldiers, including private soldiers, claiming arrears of pay, mainly for service between 1642 and 1647, are in E 315/5-6 and these may give brief service histories.

Individual soldiers who were owed arrears of pay after 1649 might be given certificates known as debentures which certified what they were owed: these debentures, secured on property which had been confiscated by Parliament, could be used to purchase such properties which could then, in turn, be sold to pay off their debts. These Certificates for the Sale of Crown Lands, in E 121, contain thousands of names of officers and men who had served in the parliamentary forces.

These records are summarised in TNA Research Guide *Civil War Soldiers 1642–1660*. More information, largely about the Scots Army of the period, can be found in Gerald Hamilton-Edwards, *In Search of Army Ancestry* (Philimore, 1977).

There are, of course, a number of websites devoted to the English Civil War, which may provide background information. I particularly liked the worksheet on soldiers' lives during the English Civil War, provided by the Royal Armouries Education Department at www.royalarmouries.org/extsite/view.jsp?sectionId=116.

Chapter 4

OFFICERS

To be an officer in the British Army in the eighteenth and nineteenth centuries, it was almost essential that you were a gentleman, for it was believed that only officers had the qualities to lead British troops in the field. Provided you were, much else could be forgiven – competence and capability hardly mattered at all. General Sir John Moore once described Lieutenant William Warre as being 'a slathering goose, who erred, as he always will do from no bad design, but from want of brain'. Despite this Warre rose to become a lieutenant general.

Age was not a bar. Officers who had fought at Waterloo were still on the strength thirty or more years after the battle. Indeed the senior commanders in the Crimea had all spent their formative years fighting the French – a fact that sometimes confused them, particularly as the French were now their allies.

The social class from which the vast majority of officers were recruited hardly changed during the eighteenth and nineteenth centuries. Over 40 per cent of colonels in 1868 came from either peerage or gentry, and another 15 per cent had fathers who themselves had been officers.

As with the other ranks, generations of the same family joined the Army and often the same regiment. One of the most successful commanders of the Second World War, Lord Wavell (himself the son of a major general) candidly wrote in his memoirs that:

> *I never felt any special inclination to a military career, but it would have taken more independence of character than I possessed at the time, to avoid it. Nearly all my relations were military. I had been brought up amongst soldiers; and my father, while professing to give me complete liberty of choice was determined that I should be a soldier. I had no particular bent towards any other profession, and took the line of least resistance.*

Most came from a rural background and an increasing proportion were educated at public school. During the Boer War nearly two-thirds of regular officers had

An officer of the 3rd Dragoons in 1845.

been to public school, 11 per cent from Eton alone. These schools were felt to develop the right character, with an emphasis on order, authority and discipline and the body and spirit through the cult of team games. Above all loyalty, to school, regiment and country was prized.

Until 1871 commissions were largely purchased, in effect buying a position or promotion in a particular regiment (but not the Royal Artillery or Royal Engineers). The Duke of Wellington stoutly defended the system, arguing that: 'It is promotion by purchase which brings into the servicemen who have some connection with the interests and fortunes of the country.'

The commission was also an investment to be sold on retirement. Depending on the regiment it might cost £450 to buy an Ensigncy (equivalent to second lieutenant) in a foot (infantry) regiment or £1,200 for the same rank in a much more fashionable guards regiment. In 1869, Hugh McCalmont paid £5,125 for a captaincy in the 9 Lancers.

It was also a gamble. If an officer died on active service, his commission was given to a deserving officer, normally the senior of the next lowest rank in the regiment. The investment was also lost if the individual rose to the rank of general. Sir James Hope Grant, for example, lost £12,000 when he was so promoted.

One of the worst abuses was the buying of commissions for boys, which allowed young men to rapidly ascend the ranks. James Wolfe, the hero of Quebec, was commissioned at age fifteen and the Duke of Wellington, himself, was a lieutenant colonel by the time he was thirty.

More commonly this allowed the rapid promotion of fashionable young blades, Thomas Budenall, the Earl of Cardigan, being perhaps the most famous example, and held back the careers of promising officers without a fortune. Henry Havelock, once grumbled that before he had become a captain, he had been 'purchased over by two fools and three sots'.

The lamentable performance of the British Army in Crimea brought public attention to the system. In 1856, a royal commission concluded that purchase was: 'vicious in principle, repugnant to the public sentiment of the present day, equally inconsistent with the honour of the military profession and the policy of the British Empire and irreconcilable with justice.' However it was not abolished until 1871 when the government paid £6,150,000 to buy back commissions from officers. It was a small sum to redeem the Army out of pawn from its officers.

Pay for an officer was low and the expenses high, particularly in the more fashionable regiments. In 1902 a Select Committee estimated that the annual expenses of dining in the mess, sport, social recreation and the constant moving of Army life presumed a private income of £100 to £150 for an infantry officer and of £600 to £700 for a cavalry officer. 'Officers,' wrote Captain W E Cairnes in a guide to the British Army, 'have lived in the 10th [Hussars] with an allowance of £500 a year, in addition to their pay, but they have rarely lasted long'.

As late the Boer War, pay was based on rates laid down during the Napoleonic Wars. An ensign or second lieutenant received 5s 6d per day, although there were a variety of allowances for linguistic skills, for servants and the like. Officers in guards and cavalry regiments were paid more.

An officer of the Royal Welch Fusiliers during the Crimean War.

Men from less wealthy families often sought service in India where expenses were much lower and consequently living standards much higher. Lord Wolseley once said that when he served in the sub-continent in the 1850s:

> *The great bulk of the young men who then went to India were socially not of a very high order. Of course, though very poor, many were sons of old officers of good families, whose poverty compelled their sons to serve in India, if serve they would in the Army. But the great bulk of those I met at Chatham, and afterwards in India and Burmah struck me, I remember, as wanting in good breeding and all seem badly educated.*

Except in wartime, life for an officer was hardly demanding. Lord Gleichen, who joined the Grenadier Guards in 1881, recalled that 'we thought ourselves badly used if (except on guard days) we did not find ourselves free by lunchtime'.

Increasingly during the nineteenth century the centre of an officer's life became the regimental mess, particularly if he was young and unmarried.

Lieutenant G E Haws of the Royal Fusiliers confessed to a friend in 1901 that: 'I am beginning to realise that I do not like soldiering, though I very much like the life – the fun of always being with a lot of nice people of one's own age. I feel I never want to leave this jolly battalion, and the thought of one day perhaps doing so fills me with dread.'

Each regimental mess had its arcane rules and traditions, faithfully passed down to new entrants. In the 2 Scottish Rifles an officer with less than three years' service was not allowed to stand on the rug in front of the anteroom fire. Smoking, in the battalion, was permitted only after the port had circulated twice.

Until 1850 officers in the Worcestershire Regiment were required to wear their swords to regiment dinners, just in case the natives attacked again as they had when a garrison in the Leeward Islands was massacred while the officers were messing in 1746. Even after this tradition was abolished the orderly officer and the captain of the week still continued to wear swords.

They also amassed collections of silver donated by guests or former members of the regiment, or purchased to commemorate an event or sporting triumph, which are now largely on display at regimental museums.

The officers had very little contact with the men of whom they were supposedly in charge. Captains had a nominal responsibility for the inspection and drill of their companies and often supervised the management of canteens, participated in courts of inquiry and courts martial and oversaw company accounts. Only the adjutant, normally a junior officer who was responsible for the day to day running of the battalion, had what might be termed a full time job.

Except in battle, life was easy. Hugh McCalmont, describing the life of a subaltern (junior officer) in the 1860s wrote: 'there was plenty going on in Dublin in those days, any amount of dances and dinners going on in the season, and real good hunting to be had fairly close to hand, while soldiering was taken pretty easy even during the drill period.'

Leave was generous. Two or three months a year was the norm. Gleichen's fellow officers in the Guards expected six months' leave a year (and sometimes got eight).

4.1 The records

Officers' records are very different from those for other ranks. They are perhaps less informative and in some ways harder to find, because there is not just one series. Until the First World War the Army did not maintain central records for officers; instead this was largely done by regimental record offices. More details are given in a TNA Research Guide *British Army: Officers' Records 1660–1913*.

As well as The National Archives it is worth approaching the appropriate regimental museum to see whether they have additional material.

4.1.1 Army lists

Brief details of Army officers have been gathered together since 1702 and published as the *Army Lists* from 1740 onwards. Since 1754 the lists have been published at least annually and generally more frequently.

The National Archives has a complete set from 1759, which is on the open shelves in the Microfilm Reading Room at Kew. In addition there are complete record sets, with manuscript amendments, of the annual lists between 1754 and 1879, and of the quarterly lists from 1879 to 1900, in series WO 65 and WO 66 respectively.

Series WO 64 contains manuscript lists of Army officers between 1702 and 1752, for which there is an index in the Research Enquiries Room. There is a TNA Research Guide *British Army Lists* which describes these books in more detail.

The British Library (BL) has a complete set of lists from 1754, together with many unofficial publications listing officers (particularly in Militia regiments) from 1642 onwards. A checklist to the holdings of the BL can be found at www.bl.uk/collections/social/srvlst1a.html and they are described in a free booklet *Service List for the Army, Navy and Air Force* which is available from the Library.

The National Army Museum, Imperial War Museum and regimental museums also have sets in various stages of completeness.

A small number of Army Lists can be consulted online at www.1837online.com for a fee. Naval and Military Press (www.naval-military-press.co.uk) has published several facsimiles and companies such as Archive CD Books (www.archivecdbooks.org) and S&N (www.sandn.net) have made the odd copy available on CD.

Most lists include regular officers with those on short service commissions but some also mention (and others are confined to) Militia, Fencible, Yeomanry, or Territorial Army officers. Depending on the type of Army List, officers who have retired on pension or half-pay may also be included. What they will not contain are details of service personnel who are not commissioned officers; that is NCOs and privates, although a few lists may contain senior NCOs.

Lists are usually arranged by regiment or command, although there is always a comprehensive surname index. Dates of promotions are also given – these dates are those when the promotion has been approved and appeared in the *London Gazette*.

There are five distinct series of Army Lists and the information they contain varies. Generally the nineteenth century ones are the most informative:

1. **Annual Army Lists** date between 1754 and 1879 and are arranged by regiment, of the regular Army only. They are indexed from 1766, but

engineer and artillery officers are only included in the index from 1803. They were continued by the Quarterly series.

2. **Monthly Army Lists** date from 1798 to June 1940 and are arranged by regiment. In addition, they include some idea of the location of each unit. Officers of colonial, Militia and territorial units are included. The lists are indexed from 1867. After July 1939 the lists were given a security classification and not published and in 1940 they were replaced by the Quarterly Army List.

3. **Quarterly Army Lists** There are two separate series of quarterly lists:

> **1879–1922** These continued the Annual series, from 1879–1922. They contain the regimental list until 1907 only. However, all include a gradation list, which is a list of officers in order of seniority, with dates of birth and promotions. From April 1881, details of officers' war service are also included. From 1909 to 1922, these details of war service appear in the January issue only. This series was replaced by the Half Yearly Army Lists in 1923.

> **July 1940–December 1950** These included a gradation list of serving officers. The January issue (but not the July) also included retired officers. From 1947 they were, in fact, issued annually in February.

4. **Half Yearly Army Lists** exist for the period between 1923 and February 1950. They replaced and took a similar form to the Quarterly Army Lists. They were issued in January and July each year and included a gradation list of serving officers. The January issue also included a list of retired officers. From 1939 they became a restricted publication. From 1947 they were issued annually in February.

5. **Army Lists and Army Gradation Lists** The Army List was revised in 1951, and now consists of three parts: part 1, a list of serving officers; part 2, a list of retired officers; part 3, a brief biography of officers called the Gradation List. Part 3 is a restricted publication, and is not available to the general public. Part 2 is published only every four years.

4.1.2 Hart's Army Lists

For the Victorian and Edwardian periods a more useful resource might be Hart's Army List, particularly as it includes short accounts of officers' war services. Lieutenant General Henry Hart started an unofficial Army list in February 1839, in part to fulfil the need for a record of officers' war services which he felt were inadequately covered in the official Army List. Hart's Army Lists cover the period between 1839 and 1915 and were issued quarterly. Of particular value are the detailed footnotes giving the active service details and campaign awards for each officer although, as much of the information was provided by the subjects themselves, it should be treated with care.

An example of this is the rather self-glorifying biographical entry for Major Gonville Bromhead (who was played by Michael Caine in the film *Zulu*) in editions published in the 1880s, which reads:

> *Major Gonville Bromhead served in the Zulu War of 1879 with the 24th Regiment, and took part as second in command in the gallant and successful defence of Rorke's Drift (mentioned in despatches, brevet major, Victoria Cross, medal with Clasp); together with Lieut J R M Chard, Royal Engineers, was awarded the VC 'for their gallant conduct at the defence of Rorke's Drift on the occasion of the attack by the Zulus on the 22nd and 23rd January 1879. The Lieutenant General commanding the troops reports that, had it not been for the fine example and excellent behaviour of these two officers under the most trying circumstances, the defence of Rorke's Drift Post would not have been conducted with that intelligence and tenacity which so essentially characterised it. The Lieutenant General adds that its success must, in a great degree, be attributed to the two young officers who exercised the chief command on the occasion in question.*

An annual volume which contained additional information was also published. Copies of the lists, and Hart's papers, are in WO 211, although you can find many volumes on the open shelves in the Microfilm Reading Room. Again, sets are available at the British Library and military museums and a number have been published as facsimiles and on CD.

4.1.3 Other lists
There are a number of other lists, which are described below, and in the British Library checklist mentioned above. TNA Library has copies of many of these volumes and they sometimes may be found in military museums and large reference libraries as well.

Details of officers granted commissions before 1727 can be found in Charles Dalton, *English Army Lists and Commission Registers, 1661–1714* (6 vols., London, 1892–1904) and *George I's Army, 1714–1727* (2 vols., London, 1910–1912). Copies of Dalton's books are available at TNA and many other reference libraries.

The Military Register, published from 1768 to 1772 and in 1779, includes Army and marine officers. *The Royal Military Calendar*, published in 1820 in ten volumes, contains service records for officers from field marshal down to major, who held the rank at the date of publication. An index by Norman

1162. Frank Sheppard Gillespie. b. 19 Oct 1889 at Killaloe, Co Clare. MB BCh BAO and LM Rot 1914; MB Dub 1918. (TC Dub). SR: Lt 28 Aug 14. Mobd 23 Sep 14. Capt 1 Apr 15. PRAC: Capt 1 May 19 (28 Feb 18). Maj 28 Aug 26. Lt Col 11 Apr 37. A/Col 4 Mar to 10 May 40 and 31 Jul to 22 Nov 40. T/Col 23 Nov 40 to 25 Jan 42 and 1 Mar 42 to 15 Apr 43. Col 16 Apr 43 (11 Apr 40). r.p. 21 Oct 48. Ceased R of O 19 Oct 49. France 1915–16. India 1919–24. BAOR 1924–25. India 1926–30. Malta 1933–37. Egypt 1937–39. CO Mil Hosp Holywood Mar–Sep 39. BEF France 1939–40: CO 4 Fd Amb. CO 35 Gen Hosp Mar–May 40. Iceland: ADMS HQ 147 Inf Bde May–Jul 40. ADMS HQ 5 Div 1940–42. USA: MLO to US Army Washington 1942–46. PSMB Devon & Cornwall Area 1947–48. 1914–15 S. BWM. VM. IGSM & cl Afghanistan 1919. 1939–45 S. DM. WM.

There are a number of biographical dictionaries which describe the careers of officers. This extract comes from List of Commissioned Medical Officers of the Army, 1660-1960 *(Wellcome Trust, 1968).*

Hurst was published in 1996. The *Calendar*, however, contains no personal information or details of officers' families.

Lists of artillery officers were published in F C Morgan's *List of Officers of the Royal Regiment of Artillery, 1716–June 1914* (3 vols., London, 1899, 1914). A similar list was compiled for the Royal Engineers by T W R Connolly: *Roll of Officers of the Corps of Royal Engineers from 1660 to 1898* (London, 1898). In addition, there is a published *List of Commissioned Medical Officers of the Army, 1660–1960* (2 vols., 1925, 1968). General staff officers and War Office staff (including civilian employees) are listed in the *War Office List* published by the War Office itself between 1861 and 1964. Copies are to be found in TNA Library.

4.2 Service records

Before the First World War there is neither a single series of service records nor a single document describing an individual's service. Instead there are ledgers containing details of promotions, plus correspondence relating to the purchase of commissions, promotions, pensions etc.

There are two main series of records of service of officers: those compiled by the War Office and those compiled in regimental record offices. An incomplete card index to names is available in the Microfilm Reading Room at Kew.

The records described below are only for officers who had retired before the end of 1913. Service records for officers who served in the First World War and later wars are described in Chapter 12 and subsequent sections.

4.2.1 War Office compilations

The War Office did not begin to keep systematic records of officers' service until the early nineteenth century, having always relied on records retained by regimental record offices.

During the nineteenth century, however, the War Office compiled five series of statements of service, which were based on returns made by officers themselves.

1809–1810 (WO 25/744-748) – arranged alphabetically and contains details of military service only.

1828 (WO 25/749-779) – refers to service completed before 1828. They are arranged alphabetically and give the age at commission, date of marriage and children's births as well as military service. Related correspondence from officers, whose surnames begin with D to R only, is in WO 25/806-807.

1829 (WO 25/780-805) – contain similar information as the second series. The Army List for 1829 is an index to it.

1847 (WO 25/808-823) – compiled by retired officers and refers to service completed before this date. They are arranged alphabetically and contain the same sort of information as the 1828 survey.

1870–1872 (WO 25/824-870) – include a few returns before 1870 and after 1872 and are arranged by year of return and then by regiment.

Occasionally these returns will include personal asides and comments. In the 1829 half-pay return, for example, when asked whether he was desirous of service Lieutenant Colonel Wentworth Serle answered: 'Most desirous and ever has been as will appear from the many applications made by him for that purpose.' His return also praised the Duke of York, when Commander-in-Chief, for his help 'in consideration of the pecuniary embarrassments which at that time did not allow me to remain in the active duties of my profession, and in some degree to give temporary relief without my being obliged to quit the service altogether'.

In 1847, Richard Henry Tolson, a major on half-pay, was one of several officers who claimed long service for his family. In a letter accompanying his return, he pointedly wrote: 'I sincerely hope, my services, as well as that of my grandfather, who died of his wounds and also my father, who had been wounded several times, and that my ancestors traced from the time of the Black Prince, were slain in battle, of which I had the honour to address you on 22nd March last, be taken into consideration.'

4.2.2 Regimental service records
Many regimental record offices kept their own records of service. Those records, which were eventually transferred to the War Office, are now in WO 76. They are arranged by regiment and cover service between 1764 and roughly 1913, although there are some entries as late as 1961. The information contained varies a great deal, but generally increased in detail during the course of the nineteenth century.

The records are available on microfilm in the Microfilm Reading Room.

Service records for officers in the Royal Garrison Regiment between 1901 and 1905 are in WO 19 and those of the Gloucester Regiment (1792–1866) in WO 67/24-27.

Lists of Royal Artillery officers between 1727 and 1751 can be found in WO 54/684, 701. Pay lists for officers from 1803 to 1871 are in WO 54/946.

Records of service for Royal Engineer officers between 1796 and 1922 are in WO 25/3913-3919. These records also include details of marriages and children.

Particulars of service and some personal information for a small number of mainly senior officers can be found in a series of selected personal files in WO 138. Among the records are the files of General Charles Gordon of Khartoum, Lord Kitchener of Khartoum, Field Marshal Bernard Montgomery, and the poet Wilfred Owen. These records are closed for seventy-five years after the last entry in the individual's file.

4.3 Commissions, appointments, transfers and promotions

Officers held their rank by virtue of a royal commission. The issue of a commission, or warrant of appointment, is likely to be recorded in several places, as well as published in the *London Gazette* and noted in the *Army List*. A small collection of original commissions between 1780 and 1874 is in WO 43/1059.

The issue of a commission, or warrant of appointment, is likely to be recorded in several places. There is a great deal of duplication in the records, and the entries sometimes give minimal information. In a number of instances, you may prefer to bypass the original documents, as much of the information they contain can be found in the *Army Lists* or published lists of officers which are described above.

The most useful source is the Commander-in-Chief's Memoranda in series WO 31, which include applications to purchase and sell commissions, between 1793 and 1870. These records are arranged chronologically by the date of appointment or promotion, usually in monthly bundles. The *Army List* will give you the year and month of appointment so it is easy to identify the appropriate bundle.

These applications may shed considerable extra light on the individual concerned. The supporting documents often contain statements of service, certificates of baptism, and letters of recommendation. There may also be correspondence, perhaps over many years, concerning an officer's career, for example over attempts to find full time employment rather than remain on a state of half-pay. As was the fashion of the period, letters are usually couched in words of extreme obsequiousness. Even so they can reveal fears about the inability to look after a growing family or the ill-health of the head of the family which compels him to find a career for a teenage son.

Warrants to issue commissions, between 1679 and 1782 are in the military entry books in the State Papers (SP 44/164-418). They are continued to 1855 in HO 51.

Between 1660 and 1803, commission books recording the award of commissions were kept by the Secretary at War and the Secretary of State for War and are in WO 25/1-121. Similar information can be found in the notification books between 1704 and 1858 in WO 4/513-520 and WO 25/122-203.

Appointments and subsequent transfers and promotions are also recorded in succession books. They were compiled retrospectively from the notification and commission books. They are in two series: by regiments (1754–1808) in WO 25/209-220; and by date (1773–1807) in WO 25/221-229. They are available on microfilm in the Microfilm Reading Room.

Commission books for Royal Engineer officers between 1755 and 1852 are in WO 54/240-247. An incomplete series of commission books for Royal Artillery officers (1740–1852) is in WO 54/237-239, 244–247, 701.

Commission books and related papers for officers in the Royal Artillery and Royal Engineers between 1740 and 1855 are in WO 54.

4.4 Purchase of commissions

The system of purchase and sale of commissions began in the reign of Charles II, and survived until 1871. Commissions could be purchased in the Guards and in the regiments of cavalry and infantry. Commissions in the Royal Engineers and Royal Artillery were only granted to those men who had been selected to attend and successfully passed, a course of instruction at the Royal Military Academy at Woolwich. Many officers purchased their commissions (up to the rank of colonel), although free commissions could be awarded by the Commander-in-Chief. If an officer died in service he forfeited his commission to the Crown. There was naturally more opportunity for promotion by merit in time of war.

Once an officer had purchased a commission then, subject to a number of provisos, he had a claim (by the seniority rule) to purchase the next highest rank, but junior officers sometimes managed to bypass this rule. Another feature of the system was that of exchanges. Officers of equal regimental rank were permitted to exchange between regiments (or battalions), subject to a number of conditions. Illegal payments for exchanges however, were a common occurrence.

The system and its abuse is explained in Anthony Bruce, *The Purchase System in the British Army, 1660–1871* (Royal Historical Society, 1980) and rather more simply in Liza Picard's *Victorian London* (Weidenfeld and Nicolson, 2005). Relevant records at The National Archives are described in a TNA Research Guide: *British Army: Officers' Commissions*.

Hart's Army Lists gives the date when an officer made his purchase. Using this information it is thus possible to discover which bundle contains details of the purchase.

Correspondence between 1704 and 1858 can be found in indexed letter books in WO 4/513-520. Of more use are the detailed papers in series WO 31 which date between 1793 and 1870. These are explained in more detail above.

Both the official Army Lists and Hart's Army Lists record when an officer sold his commission. Unfortunately an exact date is not given, but the month or quarter can be determined.

The service of every officer holding a commission on 1 November 1871 is included in the papers of the Army Purchase Commission in WO 74, together with a series of applications from officers for compensation to which certificates of service are attached. Papers and applications are indexed by regiment but not by name of applicant.

4.5 Further information and miscellaneous series

A few papers and correspondence relating to individual officers are in WO 43. Another useful but unexplored source is the War Office correspondence between 1732 and 1868 in WO 1, with indexes and letter books, containing abstracts of correspondence, in WO 2. You might find correspondence about officers in financial difficulties, seeking promotion or return to full time service from half-pay. In piece WO 2/31 for 1826, for example, there is a copy of a letter to a Lieutenant Colonel Lowther asking for repayment of £10 he was overpaid thirteen years previously in 1813. Discipline is often the subject for correspondence, for example in 1837, when there were complaints against officers in the 3 Dragoon Guards about abusing and assaulting the chief usher at the theatre in Cork.

Later records can be found in WO 32, with a few especially confidential files in WO 141.

Inspection returns in WO 27, for the period 1750 to 1857, record the presence or absence of officers from their regiments at the time of inspection and may contain a brief record of service. The absence of officers is also recorded in the monthly returns between 1759 and 1865 in WO 17. This was common because for many officers the Army was only a part time career. Matters were not helped by extremely long periods of leave.

Original submissions, and entry books of submissions, to the sovereign of recommendations for staff and senior appointments or for rewards for meritorious service between 1808 and 1914, are in WO 103. The series also includes lists of aides de camp (ADCs) appointed to serve the sovereign between 1749 and 1940.

Additions to the list of general officers receiving unattached pay, that is, pay other than from their regiment, between 1835 and 1853, are recorded in WO 25/3230-3231 generally for service in the War Office or on the staff. Staff pay books and returns, between 1782 and 1870, are in WO 25/689-743. Ledgers of the payment of unattached pay from 1814 to 1896 are in PMG 3, with registers of those receiving unattached pay between 1872 and 1880 in WO 23/66-67.

4.6 Half-pay and pensions

There are a number of overlapping sources about the provision of half-pay and pensions which may or may not provide the information you are looking for. The sources available are described in a TNA research guide: *British Army: Officers' Records 1660–1913*.

4.6.1 Half-pay

It was not until 1871 that officers were entitled to a pension as of right. Before then, men who wished to retire either sold their commissions, recouping their capital investment, or went on to half-pay. The system of half-pay was set up in 1641 for officers of reduced or disbanded regiments. In time it became essentially a retaining fee, paid to officers so long as a commission was held; thus they were, in theory if not in practice, available for future service. From 1812 there was provision for its payment to officers unfit for service.

During the nineteenth century the system became more and more heavily abused. Officers who could afford to go onto half-pay could avoid service abroad or any unwelcome posting. It was also possible to buy a commission and then go on half-pay the next day, which made officers eligible to purchase the next commission up without serving any time with the regiment.

On the other hand, William Hambley of the 3 Rifle Brigade served for eight years in Spain and France, fighting in eighteen battles and being wounded five times. He was placed on half-pay when his battalion disbanded. He remained unemployed for more than thirty-five years, when at the age of sixty-two he was recalled to duty in the Crimean War.

Officers receiving half-pay are listed in the *Army List*. There are also lists of *Reduced Officers in His Majesty's Land Forces entitled to receive half-pay in Great Britain* for 1739 and 1740 in Parliamentary Papers. The list for 1740 has been published with an index by the Society for Army Historical Research.

Registers of officers are in WO 23. A series of alphabetical registers of those in receipt of half-pay between 1858 and 1894, with name, rank, regiment, date of commencement, rate and a record of payments, are in WO 23/68-78.

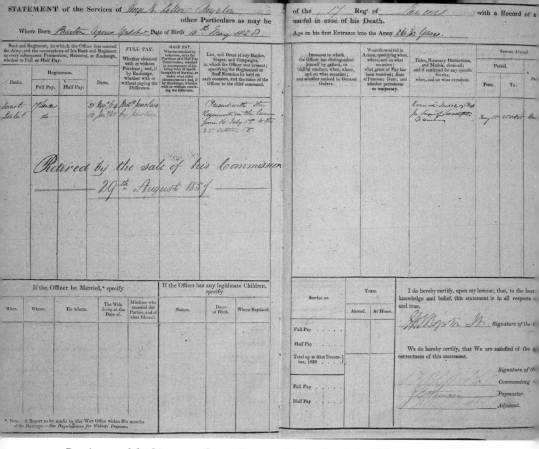

Service record for Lieutenant George Royston who served with the 17 Lancers in the Crimea. (TNA:PRO WO 76/10)

Lists of those entitled to receive half-pay between 1713 and 1809 (arranged by regiment) are also in WO 24/660–747. These are often annotated with the date of the officer's death.

Ledgers recording the issue of half-pay from 1737 to 1921 are in PMG 4. Before 1841 they are arranged by regiment and are not indexed; thereafter they are arranged alphabetically by name. Among the information noted in the ledgers are deaths, the assignment of pay, and sales of commissions, and from 1837, addresses. Later volumes also give dates of birth.

In addition WO 25/2979–3002 contains further nominal lists, for the early eighteenth century, and registers of warrants for between 1763 and 1856. Replies to a circular of 1854, with details of the fitness for service of half-pay officers, are in WO 25/3009–3012.

Registers of half-pay disbursed to officers living abroad are in WO 25/3017–3019; these cover the period 1815 to 1833 with a register of permissions granted to officers on half-pay to be abroad in WO 25/3232. These records are available on microfilm in the Microfilm Reading Room.

Claims from wounded officers for half-pay between 1812 and 1858 are contained in a series of letter books in WO 4/469-493.

4.6.2 Retired full pay

Registers of those receiving such pay, between 1872 and 1894, are included in WO 23/66-74. Earlier registers from 1830 are in WO 25/3000-3004. Ledgers of such payments between 1813 and 1896 are in PMG 3.

Registers of payments made to Army officers in reduced circumstances between 1720 and 1738 are in WO 109/55-87.

4.6.3 Pensions for wounds

A system of pensions for wounded officers was set up in 1812. Registers of those who received pensions between 1812 and 1897 are in WO 23/83-92. Correspondence on claims between 1812 and 1855 is in WO 4. Other correspondence, 1809–1857, is in WO 43, for which there is a card index in the Reference Room at Kew. Ledgers for payments between 1814 and 1920 are in PMG 9.

4.6.4 Widows' pensions

Although officers had no entitlement to a pension, provision was made from 1708 for the payment of pensions to widows of officers killed on active duty. In addition from 1818, fifteen annuities were paid to widows of officers whose annual income did not exceed £30 a year out of a fund created by the will of Colonel John Drouly.

There are several series of registers:

Reference	Dates	Notes
WO 24/804-883	1713–1829	
WO 25/325/3120-3123020-3050	1735–1769	Indexes (1748–1811) are in WO 25/3120-3123
PMG 11	1808–1920	Not April 1870–March 1882, which are in PMG 10
WO 23/105-113	1815–1892	
PMG 10	1870–1882	Continuation of PMG 11

In addition, there are several series of application papers for widows' pensions and dependents' allowances:

Reference	Dates	Notes
WO 42	1755–1908	These papers may include proofs of birth, marriage, death and probate
WO 25/3089-3197	1760–1810	Arranged alphabetically, with abstracts of the applications between 1808 and 1825 in WO 25/3073-3109. There is an index in the Reference Room at Kew. The records are available on microfilm only
WO 43	1818–1855	A few applications only

4.6.5 Children's and dependent relatives' allowances

From 1720 pensions were also paid to children and dependent relatives of officers out of the Compassionate Fund and the Royal Bounty. Registers giving the names of those placed on the Compassionate List, between 1779 and 1894, and noting the amounts they received are in WO 24/771-803 and WO 23/114-123. Ledgers recording the payments made between 1840 and 1916 are in PMG 10.

Chapter 5

OTHER RANKS: ENLISTMENT AND CONDITIONS OF SERVICE

A lone of the armies of Europe, the British Army has relied on young men voluntarily enlisting. There have only been three exceptions to this when there was conscription: during the two world wars and National Service between 1949 and 1960 at the height of the Cold War. Although a purely voluntary force had many advantages over a conscript Army in terms of morale and dedication, as the French found in 1815 and the Germans discovered in the autumn of 1914, the Army was (and remains) perennially short of men.

The physical requirements for men varied depending on whether there was a shortfall in recruits. The Army could afford to be choosier if lots of men wanted to join. In the 1880s recruits had to be between eighteen and twenty-five years of age and at least 5 feet 6 inches tall, although between 1881 and 1883 the height restriction was reduced by three inches. Recruiting sergeants did not enquire too closely about the age of recruits, particularly as they received a bounty for each man they enlisted. The Guards had their own requirements and would not accept men who were less than 5 feet 10 inches in height. But this might not always have been an advantage: tall men, Rifleman Benjamin Harris noted in his memoirs, bore the hardships of war much less well than shorter fellows.

Even with the reductions in standards, the physical condition of most young men was so poor that a high proportion was rejected. In 1900, half of all men who wanted to enlist to fight the Boers were rejected. This led to a massive outcry in the press with the result that a commission of enquiry was set up to investigate conditions in the industrial slums.

Until 1871 recruits themselves received a bounty, sometimes as much as £2, as well as the 'king's shilling'. The bounty was normally drunk away or vanished in a series of mysterious 'expenses' imposed by the recruiting sergeant.

The vast majority of recruits had been casual labourers before they enlisted. In 1870 nearly two-thirds of recruits could be so described. Agricultural labourers were preferred, being widely seen as stronger, healthier and more obedient than their slum-bred counterparts. But there was increasing demand for these men – for instance from breweries and the railways – with which the Army's poor pay and conditions could not compete.

In the 1830s over half the recruits came from Ireland and Scotland, but rural depopulation and emigration reduced the numbers of young men available. By the end of the century three-quarters of men were born in England or Wales.

Men naturally joined up for a variety of reasons. The Duke of Wellington famously claimed during the Peninsular War that:

> *people talk of their enlisting from their fine military feeling – all stuff – no such thing. Some of our men enlist from having got bastard children – some for minor offences – many more for drink; but you can hardly conceive a set brought together and it really is wonderful that we should have made them the fine fellows they are.*

Most found the Army one way to escape poverty – although ordinary soldiers were never well paid, but the pay was regular and there was normally two square meals a day. It is true as Field Marshal Sir Henry Wilson once remarked 'Jack Frost was the best recruiting sergeant'.

John Shipp, for example, spent much of his childhood in the Saxmundham parish workhouse. He was apprenticed by the overseers to a neighbouring farmer, a brutal taskmaster, from whom he escaped by enlisting as a boy (aged twelve) in the 22 (Cheshire) Regiment of Foot at Colchester in January 1797. Shipp had one of the most remarkable careers of all soldiers. As a sergeant in the 22 Foot he distinguished himself at the Battle of Bhurtpore in 1805 for which he

A private of the 7 Foot in 1770.

was awarded an ensigncy in the 65 Foot. However he found life as an officer difficult. He wrote in his autobiography that: 'The gentleman did not seem to sit easy upon me.' On his return to England he sold his commission for £250 to pay his debts. However a few years later, finding himself penniless, he re-enlisted in the 24 Light Dragoons, again rising through the ranks to become a lieutenant in the 87 Foot. His career was finished by becoming involved in a horse racing scandal, which led to a court martial which cashiered him.

Another reason was that family members had connections with the Army. Having added eighteen months to his age, Frank Richards joined the Royal Welch Fusiliers in 1901 emulating his cousin who had enjoyed his time in the Army. Others had been pupils at one of the schools set up to educate soldiers' children.

Inevitably young men were attracted by the possibility of adventure and a chance to see the world. This was always emphasised by the recruiting sergeants as they travelled the country. A recruiting poster for the 14 Light Dragoons, during the Napoleonic Wars advertised for likely candidates: 'All you who are kicking your heels behind a solitary desk with too little wages, and a pinch-gut Master – all you with too much wife, or are perplexed with obstinate and unfeeling parents ...'

In the seventeenth and eighteenth centuries colonels of regiments were given 'beating warrants', allowing them to recruit 'by beat of drum,' and recruiting parties, an officer, a sergeant and a drummer or two, marched to the regiment's recruiting area. Posters announced their arrival, and they would set up in a prominent spot where the drummers would beat the 'points of war' and the sergeant would proclaim the attractions of his regiment. George Farquhar served as an infantry officer and the characters in his play *The Recruiting Officer* are drawn from life:

> *If any gentlemen soldiers, or others, have a mind to serve Her Majesty, and pull down the French king; if any prentices have severe masters, any children have unnatural parents; if any servants have too little wages, or any husband too much wife; let them repair to the noble Sergeant Kite, at the Sign of the Raven, in this good town of Shrewsbury, and they shall receive present relief and entertainment.*

Some of those who enlisted when drunk or in a fit of anger immediately regretted their actions: in October 1807 a Lambeth wheelwright named Pearce shot himself over disagreements with his wife about his recent enlistment in the Guards, which course he had taken as a result of 'a life of idleness and extravagance'.

And inevitably some joined up out of desperation. In December 1793, at his wits' end, Samuel Taylor Coleridge fled to London. There, after spending his last money on a lottery ticket which failed to win, writing a poem on the event, and contemplating suicide, he presented himself as a volunteer for the 15 Light Dragoons under the assumed name Silas Tomkyn Comberbache. Fortunately, for him and the Army (because he had already proved a dreadful soldier) his friends quickly managed to buy him out.

There were also a small number of middle class and even aristocratic entrants – 'gentleman-rankers' – as they were known. Often they had failed to pass the entrance examinations to become officers, were students or had encountered setbacks with their professional careers. I recently did some research for a friend whose great-grandfather, Richard Cavendish, joined the 4 Hussars in 1884 for five years rising to the rank of sergeant, having been a clerk in civilian life. Family rumour suggested that Cavendish was an illegitimate son of the King of Prussia. We'll never know why he enlisted.

Because of the low opinion held of ordinary soldiers, joining up was not always universally popular with the recruit's family, particularly if they came from the respectable working classes. John Fraser recalled the reaction of his father when he enlisted in 1876:

Never have I seen a man so infuriated. To him my step was a blow from which he never thought he would recover, for it meant disgrace of the worst type. Rather he would have had me out of work for the rest of my life than earning my living in such a manner. More than that he would rather see me in my grave and he would certainly never have me in his house again in any circumstances.

The recruit was given the king's shilling, underwent a perfunctory medical examination, and was taken before a magistrate to be attested. From 1783 to 1806 and 1829 to 1847 enlistment was for life, which effectively meant until the man was too infirm to serve. Between 1806 and 1829 and 1847 and 1870 the period was reduced first to twenty-three

A trooper of the 17 Lancers in 1832.

years and then twenty-one years. In the 1870s, soldiers spent six or seven years with the colours and then passed on to the reserve, though they could extend their service to twenty-one years if they wished. (More about recruiting for Wellington's Army, and much else, can be found at http://wellington15. tripod.com/soldiers.htm).

But few did extend, because everyday life in the Army for most men was unpleasant. The historian John Laffin has suggested the paradox that many of the most noble victories in the British Army's existence had been achieved by badly treated men: 'Starved, poorly housed and woefully equipped, mercilessly worked, over-loaded and grossly underfed and underpaid, his health neglected and his private and personal needs ignored, Tommy Atkins has nevertheless done his duty and allowed his commanders to win their battles.'

Ian Knight has pointed out that to modern eyes, life in the Army had little to commend it:

> *soldiers enlisted for long periods which so exhausted them that they were fit for nothing else upon their discharge – and they could expect to spend years at a stretch in remote garrisons, a prey to all sorts of strange diseases which thrived in crowded and insanitary barracks. The daily routine was one of repetitious drill and soul-destroying boredom, which many soldiers sought to escape in cheap drink. Any lapse of discipline was punished by rigorous use of the 'cat-o'-nine-tails.*

Even so men were attracted by the opportunities for adventure, the companion-ship and the chance of a share in prize money at the end of a successful campaign.

Certainly it wasn't the pay that made soldiers enlist or kept them loyal. For most of the nineteenth century infantry privates received 1s 0d a day (1s 4d in the Royal Horse Artillery and 1s 9d as a Life Guard). This was raised in 1871 to 1s 2d, supplemented by a penny allowance for beer. Cavalry men, sappers and gunners were slightly better paid. However much of the pay was deducted for messing expenses, clothing and equipment replacement, washing, hair-cutting and barrack damages. These stoppages were much resented, particularly by new recruits who expected, at least, the uniform to be provided by the Army.

Pay could be supplemented by working as batmen (servants) to officers or by holding a Good Conduct Medal which gave an allowance of a penny a day. Corporals, sergeants and other NCOs were, of course, better paid.

As with impecunious officers, service in India, and to an extent Egypt and other colonies, was often preferred because living costs were so much cheaper. It was, for example, possible to hire native servants to do a lot of the menial work, such as cleaning kit (but never guns) and keeping the barracks tidy. Conditions generally were also better, such as the provision of separate dining halls and married quarters.

It was not until the 1790s that the first permanent barracks were built. They offered little in the way of comfort and amenities other than stout walls and roofs. A Parliamentary enquiry in the 1840s found that of 146 barracks in England and Scotland, eighty-nine lacked any proper washing facilities for men, while seventy-seven had no means of washing dirty linen. Matters in Irish barracks were even worse: 130 of 139 barracks in the survey did not provide toilets or any way that ordinary soldiers could wash themselves.

Cramped barrack rooms were foul with the stench of unwashed bodies, compounded with the odour of the dual-purpose tub which served as a urinal at night and a washtub by day. One officer at an Irish barracks during the Napoleonic Wars found that:

> *In winter the men would block up all the ventilation with old sackingand when I had to visit the rooms in the morning as Duty Officer, the atmosphere was so nauseating that I felt disinclined to touch my breakfast afterwards. Of course the soldiers had only an outside tap to wash from, which was often froze up, and even when it was not, you may imagine that few of them were bold enough to strip and swill themselves in the cold and dark of a January morning. You can smell some soldiers' feet before you enter their rooms.*

In smaller barracks eight men slept in a room, where they also ate, cleaned their arms and kit and kept personal possessions. Barrack regulations specified that each soldier should have space of 450 cubic feet: this was actually less than half allowed to a convict in prison.

Men constantly complained of being hungry and little wonder. Until 1840 breakfast was generally at 8 a.m. and dinner (lunch) at 12.30 p.m. William Cobbett, who enlisted in 1784, found young soldiers 'lay in the their berths actually crying on account of hunger. The whole week's ration was not a bit too much for one day.' A century later, A F Corbett, vividly recalled the pangs of hunger which he felt at night: 'Fortunately the heavy drinkers were light eaters, and many times I have felt along the barrack shelves and found a dry crust for my supper.'

Food was also very dull. Meals generally consisted of boiled meat, in the form of a soup or stew, eked out with bread and occasionally a piece of cheese. The average rations for most of the nineteenth century were 1½ pounds of bread, 1 pound of potatoes and ⅔ of a pound of meat (including bone and gristle) for which men were docked from their meagre pay. The quality was appalling – in the late 1870s it was nicknamed 'Harriet Lane' commemorating a young woman hacked to pieces by the murderer Henry Wainwright.

To accompany this, men drank beer, perhaps sensibly because there was unlikely to be any pure water available and tea and coffee was frowned upon.

Until 1802 the official ration was five pints of beer per day, when this was replaced by an allowance of a penny a day beer money. This was an official inducement to drunkenness for which the British soldier was already well-known. In the Peninsular Campaign, a sergeant of the 43 Foot remembered:

It was no infrequent thing to see a long string of mules carrying drunken soldiers to prevent them falling into the hands of the enemy. The new wine was in tanks particularly about Valladolid and the men ran mad. I remember seeing a soldier lying fully accoutred with his knap-sack in a large tank, he had either fallen in or had been pushed in by comrades. I saw a Dragoon fire his pistol into a large vat contain thousands of gallons; in a few minutes we were up to our knees in wine, fighting like tigers for it.

Drunkenness was undoubtedly exaggerated by the fact that so few were allowed to marry. Less than 10 per cent of men who wanted to were allowed to marry 'on the strength' as the phrase went, that is with permission of the regimental authorities. Small separation allowances and a grant of one year's pay, granted to widows of men killed in action, were introduced in the 1870s and 1880s. This was of no help to the majority of men who married 'off the strength' and whose wives in consequence received no allowance at all.

Conditions began to improve after the Crimean War, which had exposed so much that was wrong with the Army. Barracks were rebuilt with proper ablutions. The basic meat and bread ration was supplemented by vegetables, spices, butter and especially tea, which was often supplied by the regiments at cost. Regimental institutions and privately run soldiers' homes and reading rooms provided quieter alternatives to the bar and the pub. Basic education was provided for those who wanted to get on: you had to pass various certificates to be promoted to a non-commissioned officer. Even so it was estimated in 1888 that 60 per cent of the ranks had failed to attain even the lowest Army education certificate, which involved simple reading and the completion of a few sums which it was thought would not present a problem to an eight-year-old child.

The Records

5.1 Service records

5.1.1 Soldiers' documents

The most important records are attestation and discharge papers, often referred to as soldiers' documents, in series WO 97 which covers the period between about 1760 and 1913. Some papers between 1785 and 1813 are in WO 121. They

A page from the Soldiers' Documents for Gunner John Hardy, Royal Artillery. He was discharged because of his bad conduct. (TNA:PRO WO 97/1797) are described in more detail in a Research Guide *British Army: Soldiers' Discharge Papers, 1760–1913.*

They are almost always referred to as 'service' records. But as the medals expert John Sly reminds us: 'They are not, they are pension records, specifically Royal Hospital, Chelsea, pension records. They are copies of documents that were given to a largely illiterate soldiery, specifying the man's name, trade and place of birth, as well as his physical description, forming a proof of his identity and entitlement to draw his pension if he could not sign his own name.' More about pension records can be found in Chapter 9.

Once this is appreciated it immediately becomes obvious why so few individual records survive: only a relatively small proportion of soldiers were pensioned, either for disability or long service. In particular if a soldier died in service, or deserted and so did not complete his term of enlistment, it is unlikely that there will be a record for him. Researchers should not forget the records for the Royal Hospital in Dublin, Kilmainham, which was operating in tandem with Chelsea at this time, and it was not specifically an institution for Irishmen or soldiers from 'Irish' regiments. It is probably because the Kilmainham records are not filed or indexed in the same way as the Chelsea records that they are so often ignored.

Before 1883 these service records are normally only for men who were discharged and received a pension. From 1883 to 1913 the series also includes soldiers who were discharged for other reasons, such as termination of limited engagements or discharge by purchase.

The types of document that have most commonly survived are, although it is unusual to find them all in a single file:

- discharge forms, which were issued when a soldier left the regiment. The purpose, initially at least, seems to have been to have been proof for the poor law authorities that the individual was not a vagrant as he passed through the area on his way home or in search of work.
- attestation forms, which are the documents signed by the new recruit. They will tell you how old he was at enlistment, where he enlisted and his trade before he joined up. There may also be details of next of kin.
- the proceedings of a regimental board and record of service, which was a more detailed record of discharge.
- supporting correspondence. Occasionally there might be scribbled notes on the application itself.
- questionnaires of past service, which an applicant for a pension completed if others documents had not survived.
- affidavits, which out-pensioners outside London made every quarter to state that they were not drawing on other public funds.

Except for the earliest documents, where the level of detail is limited, the documents give information about age, physical appearance, birthplace and trade or occupation on enlistment in the Army. They also include a record of service, including any decorations awarded, promotions and reductions in rank, crimes and punishments, and the reason for the discharge to pension. In some cases, the place of residence after discharge and date of death are also given.

These documents are arranged by discharge date and then by regiment or by surname. The order within individual piece numbers is roughly alphabetical. If you can't find your man then it is also worth looking though the misfiled records to see if he appears there.

With the exception of some records between 1883 and 1913, the records have been microfilmed and are available in the Microfilm Reading Room at Kew. The documents fall into four series:

1760–1854 These documents are arranged alphabetically by name within regiments. Fortunately there is a comprehensive index which is available on TNA's online catalogue (just type in the name of the person you are researching). This index also includes records found in WO 121. There seem to be relatively few for men who enlisted before 1792.

1855–1872 These are again arranged alphabetically by name within a regiment, and it is vital to know the regiment in which a man served. At the time of writing these records are being indexed and the results will be added to TNA's online catalogue in due course.

1873–1882 These are arranged alphabetically by name of soldier by cavalry, artillery, infantry and miscellaneous corps.

1883–1913 The documents are arranged in surname order. Details of next of kin, wife and children are given.

5.1.2 Additional series of service records

If you can't find your man in WO 97 and think that he survived to receive a pension, there are several other places for you to look.

WO 116 and WO 117 are pension books which contain similar information, but are split into disability and long-service respectively, and are indexed chronologically by date of admission to Chelsea pension. WO 120 contains pension books arranged regimentally by order of precedence, then chronologically by date of admission to Chelsea pension.

WO 118 and WO 119 are records for Kilmainham Hospital in Dublin, WO 118 being arranged chronologically by date of admission to pension, giving a numerical key to WO 119, which is the equivalent of WO 97, but in bound book form rather than loose in boxes. This can be a long process if you do not know the date of admission to pension, and you may need to use the muster books (WO 12) to get a date, but it is very rewarding if you find the man you want.

General registers of discharges from 1871 to 1884 are in WO 121/223-238. Registers of men discharged without pension between 1884 and 1887 are in WO 121/239-257. Certificates of service of soldiers awarded deferred pensions, 1838–1896, are in WO 131.

Two series of returns of service of NCOs and men survive from the early nineteenth century. The first series contains statements of periods of service and of liability to service abroad on 24 June 1806 and are in WO 25/871-1120. The second series contains returns of NCOs and men, not known to be dead or totally disqualified for military service, who had been discharged between 1783 and 1810 (WO 25/1121-1131). Both series are arranged by regiment and then alphabetically by surname.

Records of service of soldiers in the Royal Horse Artillery between 1803 and 1863 are in WO 69. They include attestation papers and show name, age, description, place of birth, trade, and dates of service, of promotions, of marriage and of discharge or death. These records are arranged under the unit in which the individual last served, which can be ascertained from indexes and posting books in WO 69/779-782, 801–839.

There is also an incomplete series of registers of deceased, discharged or deserted men in the Royal Artillery (1772–1774, 1816–1873) in WO 69/583-597, 644–647, arranged by artillery regiment. A number of miscellaneous pay lists and other records of the Royal Artillery, 1692–1876, are in WO 54/672-755.

5.1.3 Description books

Description books give a description of each soldier (whether or not he lived to receive a pension), his age, place of birth, trade and service. It is extraordinary to think that you can find the colour of your soldier's eyes and hair as well as physical deformities; information which would be impossible to discover for any civilian ancestors.

There are two main series. The regimental description and succession books are in WO 25/266-688, covering the period between 1778 and 1878, but for most regiments there are volumes only for the first half of the nineteenth century. Entries in them are arranged alphabetically, by date of enlistment or service number.

Similar books for regimental depots (1768 to 1913) are in WO 67. Helpfully they are listed by regiment rather than the depot itself, although records only survive for the Scots Guards; 8, 17, 20, 34, 29, 40, 74 and 84 Foot; Durham Light Infantry; Gloucestershire Regiment; Royal Warwickshire Regiment; South Wales Borderers; 10 Royal Hussars; Military Train and Commissariat

Staff Corps. These records tend to be more complete, because they listed people as they joined, although they do not include men who transferred between regiments.

Description books for Royal Artillery battalions between 1749 and 1859 are in WO 54/260–309, and depots between 1773 and 1874 in WO 69/74–80. Books for the Royal Irish Artillery, 1756–1774, are in WO 69/620.

5.2 Pay Lists and Muster Rolls

Muster rolls and pay lists provide a comprehensive means of establishing an ancestor's date of enlistment, his movements throughout the world and his date of discharge or death. The first entry may show his age on enlistment, as well as the place where he enlisted. An entry on the form 'men becoming non-effective', sometimes found at the end of each quarter's muster, shows the birthplace, trade and date of enlistment of any soldier discharged or dying during the quarter. Between 1868 and 1883, at the end of each muster (or the beginning for regiments stationed in India) is a marriage roll, which lists wives and children for whom married quarters were provided.

The main series of muster books and pay lists are arranged by regiment and are bound in volumes covering a period of twelve months. They were compiled monthly. They are in the following separate series for:

Unit	Dates	Series reference
Artillery	1708–1878	WO 10
Engineers	1816–1878	WO 11
Foreign Legions	1854–1856	WO 15
General	1732–1878	WO 12
Militia and Volunteers	1780–1878	WO 13
New series	1878–1898	WO 16
Scutari Depot	1854–1856	WO 14

WO 12 is the most important of these as it includes household troops, cavalry, Guards, regular infantry, special regiments and corps, colonial troops, various foreign legions and regiments, and regimental, brigade and other depots. WO 14 and WO 15 relate to troops engaged in the Crimean War. Except for a few units who served in the sub-continent during the eighteenth century, WO 10 does contain any records for artillery units who served in India.

WO 16 includes records of the infantry, artillery and engineers. From 1888 they are company muster rolls only, arranged chiefly by regimental districts. The *Army Lists* contain indexes to regiments with their regimental district numbers.

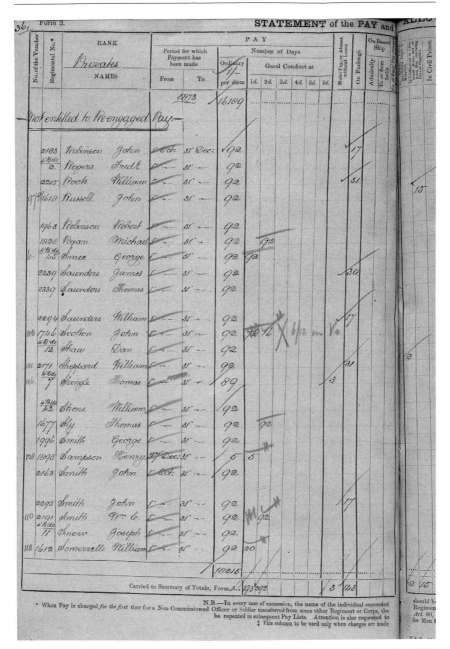

Muster roll for the 1st Battalion, 19 Foot showing money paid to privates for December 1873.
(TNA:PRO WO 12/3651)

5.3 Miscellaneous Records

It is occasionally possible to find correspondence about individual soldiers, who were perhaps deserters or sought transfer to another regiment, in the War Office correspondence in WO 1, between 1754 and 1868, with letter books and indexes in WO 4 and WO 2.

It was possible to buy discharge from the Army for the sum of £20 or less, depending on the length of service. For most men this was an enormous amount which neither they nor their families could afford and a far cheaper and easier option was just to desert. Registers, arranged by regiment, for the period between 1817 and 1870 are in WO 25/3845-3868.

5.4 Indexes

There are a number of indexes compiled by private individuals or small groups of enthusiasts of soldiers in various campaigns. They include:

5.4.1 Napoleonic Wars

Barbara Chambers maintains a number of indexes of men who served in the Napoleonic Wars. More details can be found on her website at http://members.aol.com/BJCham2909/Napoleonic_Wars.html (which also has a useful history of the period) or you can write to 39 Chatterton, Letchworth Garden City, SG6 2JY. There is a charge for this service.

5.4.2 The Great Crimean War Index

A long term researcher in the field, Brian Oldham, has compiled a detailed index to all the officers and men who served in the Crimea. It is based on the medal rolls in WO 100, but it is wider than this record. As Major Oldham has found this is a deeply flawed source, so he has been through the muster rolls and related records for the regiments who served in the Crimea. He has also included soldiers who died onboard ship on their way to the Black Sea.

The available information varies greatly. At a minimum you will find the regiment a man served with, which will enable you to check other records at Kew. Often there will be a link to the medal rolls, which will tell whether a man was included and to which clasps (Alma, Balaklava, Inkerman and Sebastopol) he was entitled.

Information in the index can be obtained from the author. More information can be obtained at www.crimeanwar.info. Alternatively you can email archive@crimeanwar.info or write to The Great Crimean War Index, 21 Malting Close, Stoke Goldington, MK16 8NX. At the time of writing a basic search to determine which regiment a man was with costs £5 and a full search £10.

There is an article about the index in issue 13 of *Ancestors Magazine* and a more general article about the Crimean War by Brian Oldham in issue 10.

5.4.3 The Rorke's Drift Research Project.

In partnership with a number of British military museums, such as the Royal Regiment of Wales' Regimental Museum in Brecon and the Army Medical Services Museum at Aldershot, the Rorke's Drift Research Project is attempting to establish the definitive database of information relating to the battle itself and to the lives of the individual participants.

It is led by Lee Stevenson who has spent the last twenty years researching the defence of Rorke's Drift and the Army careers of the men who were there. At present he is completing a major search of the Army muster and pay rolls for all of the defenders. Team members are also trying to establish contact with any descendants of those men who took part in the battle, in order to share information and help update our knowledge of their ancestors.

If you can help, please email Lee Stevenson at RDRP@fsmail.net or write to RDRP, PO Box 270, Chichester, West Sussex.

Further reading

There are a number of books on the experiences of soldiers and officers in the eighteenth and nineteenth century British Army, but among them are:

JM Brereton, *The British Soldier: A Social History from 1661 to the Present Day* (Bodley Head, 1986)

Byron Farwell, *For Queen and Country: a Social History of the Victorian and Edwardian Army* (Allen Lane, 1981)

Eileen Hathaway, *A Dorset Rifleman: the Recollections of Private Benjamin Harris* (Shinglepicker, 1995)

Richard Holmes, *Redcoat: The British Soldier in the Age of Horse and Musket* (HarperCollins, 2001)

— (ed), *Sahib: The British Soldier in India 1750–1914* (HarperCollins, 2005)

John Laffin, *Tommy Atkins: The Story of the English Soldier* (Sutton, 2004)

Edward Spiers, *The Late Victorian Army 1868–1902* (Manchester University Press, 1992)

The National Army Museum has a booklist of memoirs and accounts of life of ordinary soldiers at www.national-army-museum.ac.uk/pdf/readinglist/Social-Life-of-Other-Ranks.pdf.

There are a few books on researching individuals:

Barbara Chambers, *Army Research Other Ranks During The Napoleonic War* (available from the author at 39 Chatterton, Letchworth Garden City, SG6 2JY, price £5 including postage and packing.)

Steve Dymond, *Researching British Military Medals* (Crowood Press, 1999)

Gerald Hamilton-Edwards, *In Search of Army Ancestry* (Phillimore, 1977) – very dated, but useful for officers.

William Spencer and Simon Fowler, *Army Records for Family Historians* (2nd edition, Public Record Office, 1998)

William Spencer, *Medals: the Researcher's Guide* (The National Archives, 2006)

— *Records of the Militia and Volunteer Forces, 1757–1945* (Public Record Office, 1997)

Phil Tomaselli, *The Boer War* (Federation of Family History Societies, 2006)

— *The Crimean War* (Federation of Family History Societies, 2006)

— *The Zulu War* (Federation of Family History Societies, 2006)

Chapter 6

MEDALS, HONOURS AND AWARDS

Some families are lucky to be able to treasure the medals awarded to an ancestor. Even if those for soldiers on your family tree have long since disappeared, medal rolls and other records can be informative, particularly as almost every soldier from the mid-nineteenth century onward was awarded one or more medals.

Medals in their present form are a nineteenth century innovation. During the eighteenth century special medals in gold or silver were sometimes struck to commemorate great victories, but they were privately minted and usually given only to senior officers. More details are given in William Spencer, *Medals: the Researcher's Guide* (The National Archives, 2006)

There are three types of medal: campaign, gallantry, and long service and good conduct.

6.1 Campaign medals
The names of the men awarded medals for campaigns between 1815 and roughly 1914 are recorded in medal rolls. Most are in series WO 100 at Kew. They are arranged first by regiment, then rank, then name. For most medals the only information given is the recipient's number and a note of the bars to which he was entitled. Both the Queen's South Africa Medal and the King's South Africa Medal rolls (issued for service during the Boer War), however, do give more information about the service of individuals. These records are only available on microfilm. In addition, there are various files dealing with recommendations for awards, mainly for small colonial campaigns.

The rolls for many campaigns have been published, often with additional information about individual recipients, so they are worth seeking out. TNA Library has a good collection. Regimental and other military museums, and larger reference libraries should also have copies.

The first campaign medal was the Waterloo Medal which was granted to soldiers who fought at the battle on 18 June 1815 and the preliminary skirmishes on previous days. It was also the first campaign medal awarded to the next of kin of men killed in action. Lastly, it was the first on which the name of the recipient was impressed around the edge by machine. Rolls for this are in MINT 16/112 and WO 100/14-15. The recipients of the medals are listed in *Waterloo Medal Roll Compiled from the Muster Rolls* (Naval and Military Press, 1992).

It was not until 1847 that men who served during the Peninsular War and the War of 1812 were honoured with a campaign medal – the Military General Service Medal. Survivors had to claim for the medal: 25,600 did so. Twenty-nine clasps were authorised, which were to be worn on the medal ribbon, commemorating battles and skirmishes from Egypt in 1801 to Toulouse in 1814. The medals for this are in WO 100/1-13. Recipients are listed in ALT Mullen,

An extract from A L T Mullen, The Military General Service Roll 1793-1814 *(London Stamp Exchange, 1990) showing the information which printed medal rolls can contain.*

82nd, or The Prince of Wales's Volunteers, Regiment

Agnew, T.R., Ensign	2 – Tal Vitt	Leg amputated. Capt 22/6/15. Later 2nd RVB. Deputy Storekeeper, Tipner
Begble, Thomas, Capt	1 – Talavera	Toulouse in Foster. HP 2/11/32. Glendining 1953, Spink 1987, 1 bar Talavera
Bertles, Henry, Bde Maj	3 – Rol Vim Cor	Staff Officers List
Boyd, John, Maj	6 – R V C Barrosa V P	Later 91st Foot. Glendining 1952
Brown, John, Capt AAG	3 – Rol Vim Cor	See Staff Officers List
Carew, Robert, Capt	1 – Talavera	
Davies, Edward, Lt	3 – T Vitt Nivelle	Sotheby 1980. London Stamp Exchange 1989 with another medal named to Davis
De Rency, Geo. Webb, Lt, Maj	1 – Vittoria	Later 4th Dgns. Leg amputated. Bk Mr Dundee
Drummond, John M., Lt	3 – Vitt P Orthes	Capt HP 4/12/23
Fitzgerald, Wm. Ed., Capt	4 – V C V P	
Fraser, David, Lt	2 – Nivelle Orthes	HP 25/3/17
French, E.F., Lt	2 – Nivelle Orthes	
Garnett, James S., Lt	3 – Vitt Nivelle O	
Holdsworth, Samuel, Lt and Adjt	3 – Vitt Nivelle O	Paymr 22/9/25, HP 27/8/41
Lacy, Saml. Walter, Lt	2 – Vitt Pyr	Later 10th Foot. HP 24/6/24
Mason, William, Lt	2 – Nivelle Orthes	Severely wounded at Niagara. Capt HP 5/2/47
Pigott, H., Lt	2 – Tal Vitt	
Pratt, Charles, Lt	3 – R V C	Later 96th Foot. Maj HP -/3/14
Proctor, H.A., Capt	1 – Barrosa	
Pynn, Henry, Capt	5 – R V Busaco V T	Capt 82nd Foot, Maj 18th Portuguese. British Officers in Portuguese Army list
Raines, J.R., Lt	2 – Vim Cor	Later 95th and 48th Foot
Starkie, W.W, Lt	6 – R V V P Nivelle O	
Sterne, William, Lt	4 – R V C P	Later 57th Foot

The Military General Service Roll 1793–1814 (London Stamp Exchange, 1990). An article on the medal appears in the April 2006 (no. 44) issue of *Ancestors Magazine*.

The references for rolls for other nineteenth century campaign medals are:

Abyssinia (1867–1868)	WO 100/43
Afghanistan (1878–1880)	WO 100/51-54
Africa (various) (1892–1903)	WO 100/76-77, 79, 83, 90–93
Ashanti etc (1857–1896)	WO 100/42, 44
Burma (1885–1889)	WO 100/69-70 73
China (1842–1860)	WO 100/40-41
China (1900)	WO 100/94-99
Egypt (1878–1882)	WO 100/54-61
India (1803–1826)	WO 100/13
India (1849–1903)	WO 100/20-21
India (1888–1895)	WO 100/74-75, 78
India (1897–1898)	WO 100/84-89
India and Perak (1875–1876)	WO 100/45
Indian Mutiny (1857)	WO 100/35-39
Kurdistan (1925)	WO 32/3564
New Zealand (1845–1866)	WO 100/18
New Zealand (1861, 1863)	WO 32/8258, 8270
Nubia, Sudan (1926–1927)	WO 32/3537
Punjab (1848–1849)	WO 100/13
Queen's Mediterranean (1899–1902)	WO 100/368
Rhodesia (1898)	WO 32/7840, 7842–7843
Sierra Leone (1898)	WO 32/7629, 7632, 7635
Small campaigns (1845–1879)	WO 100/19
Somaliland (1901–1904)	WO 100/100-101
Somaliland (1903–1904)	WO 32/8428, 8440
South Africa (1834–1853)	WO 100/17
South Africa (1877–1879)	WO 100/46-50
South Africa (1878–1879)	WO 32/7682, 7764
South Africa (1899–1903)	WO 32/7960, WO 108/136-179
South Africa (King's) (1901–1902)	WO 100/302-367, 369–370
South Africa (Queen's) (1899–1901)	WO 100/112-301, 371, 381–383
Sudan (1884–1886, 1896–1898)	WO 32/3539
Sudan (1884–1889)	WO 100/62-68, 71–72
Sudan (1896–1898)	WO 100/80-82
Tsingtao, China (1914–1915)	WO 32/4996B

6.2 Long service and good conduct medals

In 1833 a Long Service and Good Conduct Medal was instituted for soldiers who had served eighteen years in the Army, without a major blemish to their character. Medal rolls for this medal between 1831 and 1953 are in WO 102. The series also contains some rolls for medals issued to men serving in Militia and colonial forces.

In 1846 a Meritorious Service Medal was authorised for sergeants and warrant officers who had performed good service other than in battle. Awards for meritorious service between 1846 and 1919 are in WO 101. A register of annuities paid to recipients of the meritorious or long service awards between 1846 and 1879, is in WO 23/84. Rolls for the Volunteer Officers' Decoration, 1892–1932, are in WO 330.

6.3 Gallantry medals

Gallantry medals were awarded for a specific act of heroism. The Victoria Cross is, of course, both the highest gallantry award and the most famous. The first award for gallantry was actually the Distinguished Conduct Medal, which was established in December 1854 to reward other ranks for distinguished service in the Crimea.

The awards of gallantry medals are recorded (gazetted in the jargon) in the *London Gazette*. Initially citations, that is a description of how the medal came to be won, was also printed, although by mid-1916 only citations for the Victoria Cross and the higher gallantry medals were generally printed. Copies of the *Gazette* are widely available, including on microfilm in series ZJ 1 at TNA. Indexes to medals can be found in the Microfilm Reading Room. Copies of the *Gazette* for the twentieth century have been digitised and are available online at www.gazettes-online.co.uk. For some reason pages on this site often appear blank – the answer is to scroll down to what you are looking for.

In 1856 the Victoria Cross was instituted and the first awards were announced in the *London Gazette* of 24 February 1857 for heroes of the Crimean War. It quickly became the most coveted of all British gallantry medals. Registers of the award of the Victoria Cross and submissions to the sovereign, 1856–1953, are in WO 98. For an example see figure 13 on page 77. A list of recipients of the medal, 1856–1946, is in CAB 106/320. Files about the award of the medal to individuals between 1856 and 1957 are in WO 32 (code 50M). Citations for the Victoria Cross during the Second World War are in CAB 106/312.

There are numerous biographies of the 1,355 VC winners. The best place to find out more is on a website www.victoriacross.net. A recent book by John Glanfield, *Bravest of the Brave: The Story of the Victoria Cross* (Alan Sutton, 2005) describes their achievements.

War Office, May 2, 1879.

THE Queen has been graciously pleased to signify Her intention to confer the decoration of the Victoria Cross on the undermentioned Officers and Soldiers of Her Majesty's Army, whose claims have been submitted for Her Majesty's approval, for their gallant conduct in the defence of Rorke's Drift, on the occasion of the attack by the Zulus, as recorded against their names, viz. :—

Regiment.	Names.	Acts of Courage for which recommended.
Royal Engineers ... 2nd Battalion 24th Regiment	Lieutenant (now Captain and Brevet Major) J. R. M. Chard Lieutenant (now Captain and Brevet Major) G. Bromhead	For their gallant conduct at the defence of Rorke's Drift, on the occasion of the attack by the Zulus on the 22nd and 23rd January, 1879. The Lieutenant-General commanding the troops reports that, had it not been for the fine example and excellent behaviour of these two Officers under the most trying circumstances, the defence of Rorke's Drift post would not have been conducted with that intelligence and tenacity which so essentially characterised it. The Lieutenant-General adds, that its success must, in a great degree, be attributable to the two young Officers who exercised the Chief Command on the occasion in question.
2nd Battalion 24th Regiment	Private John Williams ...	Private John Williams was posted with Private Joseph Williams, and Private William Horrigan, 1st Battalion 24th Regiment, in a distant room of the hospital, which they held for more than an hour, so long as they had a round of ammunition left : as communication was for the time cut off, the Zulus were enabled to advance and burst open the door ; they dragged out Private Joseph Williams and two of the patients, and assagaied them. Whilst the Zulus were occupied with the slaughter of these men a lull took place, during which Private John Williams, who, with two patients, were the only men now left alive in this ward, succeeded in knocking a hole in the partition, and in taking the two patients into the next ward, where he found Private Hook.
2nd Battalion 24th Regiment	Private Henry Hook ...	These two men together, one man working whilst the other fought and held the enemy at bay with his bayonet, broke through three more partitions, and were thus enabled to bring eight patients through a small window into the inner line of defence.
2nd Battalion 24th Regiment	Private William Jones and Private Robert Jones	In another ward, facing the hill, Private William Jones and Private Robert Jones defended the post to the last, until six out of the seven patients it contained had been removed. The seventh, Sergeant Maxfield, 2nd Battalion 24th Regiment, was delirious from fever. Although they had previously dressed him, they were unable to induce him to move. When Private Robert Jones returned to endeavour to carry him away, he found him being stabbed by the Zulus as he lay on his bed.
2nd Battalion 24th Regiment	Corporal William Allen and Private Frederick Hitch	It was chiefly due to the courageous conduct of these men that communication with the hospital was kept up at all. Holding together at all costs a most dangerous post, raked in reverse by the enemy's fire from the hill, they were both severely wounded, but, their determined conduct enabled the patients to be withdrawn from the hospital, and when incapacitated by their wounds from fighting, they continued, as soon as their wounds had been dressed, to serve out ammunition to their comrades during the night.

MEMORANDUM.

Lieutenant Melville, of the 1st Battalion 24th Foot, on account of the gallant efforts made by him to save the Queen's Colour of his Regiment after the disaster at Isandlwana, and also Lieutenant Coghill, 1st Battalion 24th Foot, on account of his heroic conduct in endeavouring to save his brother officer's life, would have been recommended to Her Majesty for the Victoria Cross had they survived.

Register for the Victoria Cross showing the award of the medal to survivors of the Battle of Rorke's Drift. Recently digitised, the register is now available online on The National Archives' website. (TNA:PRO WO 98/4). Inset The Victoria Cross.

The Victoria Cross was the only medal for gallantry which could be awarded to officers until the Distinguished Service Order (DSO) was instituted in 1886 although the CB (Companion of the Bath) was frequently given to field officers for distinguished services.

Registers for the DSO and the Military Cross (MC) are in WO 389 and WO 390, with some files about the award of the Order in WO 32 (code 52C). Registers of the award of the CB, 1815–1894, are in WO 104. A name index for those who were awarded the DSO (and Military Cross) between 1915 and 1938 is in WO 389/9-24.

Submissions to the Sovereign for the award of the Distinguished Conduct Medal are in WO 146 and often contain citations. A few files about the award of the medal to individuals are in WO 32 (code 50S). Individual recipients are listed in P. E. Abbott, *Recipients of the Distinguished Conduct Medal, 1855–1909* (London, 1975).

For the First World War, lists of recipients of the DCM are in WO 391, with a card index of holders in the Microfilm Reading Room at Kew. Recipients are

Medal roll for the Queen's South Africa Medal, 1899–1901 for the Imperial Yeomanry. (TNA:PRO WO 100/120)

Regimental Number	Rank	Name	Remarks
25002	Corp.	Baker H	Invalided Home
29268	Tpr.	Baldwin A.J.N.	Invalided H.
25003	Lce Corp	Becker C.J.	
24098	"	Bevan N	
22760	Tpr.	Beynon R.R.	Invalided H.
21178	Corp.	Bignell A.J.	
26982	Tpr.	Beveridge J.	
21177	"	Barton N.J.	
29274	"	Blount J.N.	Discharged Claud.
30946	Tpr.	Bourton C.N.	
21326	"	Bramhall J.	
30955	"	Bridges B.N.	
22763	"	Brokerton S.R.	
31856	"	Burrows N	Invalided H.
27626	Lce Corp	Butler N.G.	
24417	Tpr.	Brooker N.R.	Invalided H.
25004	Sergt.	Barber E.A.	Discharged Cl.
20616	Bugler	Crowther G.	Invalided H.
21322	Lce Corp	Chatfield J.N.	
30618	Tpr.	Clark H.C.	Invalided H.

also listed in R. W. Walker, *Recipients of the Distinguished Conduct Medal, 1914–1920* (Midland Records, 1981). Most recommendations for the VC, DSO, MC, DCM and MM (and a very few Mentions in Despatches) for the Second World War (and afterwards) are in WO 373.

More about the award of gallantry medals for the two world wars can be found in Chapters 15 and 16.

6.4 Mentions in Despatches

Medals were not the only honour available. A man could be mentioned in despatches. This was one of the oldest ways of recognising meritorious service or gallant behaviour. In reports after battles and campaigns, commanding generals would single out those who had been brought to their attention, sometimes merely listing their names, but occasionally including a brief description of their services. Before 1843 only officers were mentioned. There was no distinctive decoration or even a certificate for those singled out, but the names were published in the *London Gazette* and, for officers, mentions in despatches were listed in the Army List.

A few files about individuals who were mentioned in despatches are in WO 32 (code 51). A list of people mentioned in despatches during the South African War, 1901, is in WO 108/142.

Further Reading

There are many books about medals, some of which are fairly technical, as well as a number of websites. The best introduction is three attractive books by Peter Duckers: *British Campaign Medals 1815–1914*, *British Campaign Medals 1914–2000* and *British Gallantry Awards 1855–2000*, published by Shire Publications.

Another useful book is William Spencer's *Medal Records: a Researcher's Guide* (TNA, 2006). The National Archives has also produced a number of Research Guides, which can be downloaded from the TNA website, including *British Armed Services: Campaign Medals, and other Service Medals*, *British Armed Services: Gallantry Medals* and *British Armed Services: Gallantry Medals, Further Information*.

The most comprehensive guide, however, is EC Joslin, AR Litherland and BT Simpkin, *British Battles and Medals* (6th edition, Spinks, 1988). Earlier editions were edited by Major LL Gordon. Also useful is the annual *Medals Yearbook* (Token Publishing), which includes guides to how much medals are worth as well as comprehensive listings.

Chapter 7

CASUALTY RECORDS

To modern eyes one of the worst aspects of a soldier's life in the eighteenth and nineteenth centuries was the lack of medical care and, perhaps worse, the lack of interest of the authorities in providing it until shamed into doing so during the Crimean War. Even so, it was not until the First World War that more men were killed in action or died of their wounds than from generally preventable sickness.

In part this was because medicine in general was very primitive and the treatment (or lack of it) meted out to soldiers was not very different from that available to civilians as a whole. However, Army medical men were rarely as skilled as their civilian counterparts and they were often reluctant to adopt new techniques. After the Boer War, for example, the King's doctor Sir Frederick Treves who had volunteered as a surgeon in South Africa, told a commission of enquiry that the Royal Army Medical Corps was still using surgical 'instruments I should have thought would only be found in museums'.

The treatment of the sick and wounded during the Peninsular and Crimean Wars was really little different from that which had gone on during the campaigns of the eighteenth century. Men endured wounds with fortitude, leading Professor Mervyn Singer from London's Institute of Intensive Care Medicine to claim recently that the impressive survival statistics of injured soldiers serve as a reminder of how we underestimate the human body's ability to heal itself under the most extreme conditions. He says that the fifty-two privates in the 13 Light Dragoons wounded by sabre, gunfire and cannon injuries at Waterloo, only two subsequently died.

Often a man's sufferings only began when he arrived at a temporary hospital established in a barn or church. Here surgeons in shirt sleeves wielded saw and knife and probed with more enthusiasm than skill, amid a foul mess of blood, rags and dirt. Fewer than half of their patients could expect to survive gangrene,

Inspecting the dead after the Battle of Inkerman.

loss of blood or tetanus. Operations were conducted without anaesthetic and the only disinfectant available was vinegar.

Many others succumbed to shock, dehydration and the loss of blood – much of it the result of artificial bleeding which was the current medical fashion. Lieutenant George Simmons, of the Rifle Brigade, was struck by an enemy ball in the chest at Waterloo. As well as removing the ball, the surgeons bled a quart of blood. In great pain he was sent to Brussels (ten miles away) on a horse where he was billeted with friends. It was not until a month later that he began to recover after his wound burst, releasing an enormous amount of pus.

It is little wonder that even severely wounded men preferred to avoid the surgeon's knife. During the Peninsular War, Private William Dougald was hit in the thigh by three balls in the space of five minutes, and although the wounds were severe, he refused to seek medical aid. A few days later, now almost lame, his regiment found themselves in another battle with the French. Ordered to the rear for treatment, Dougald said that he would rather die than leave his comrades. In fifteen minutes he had his wish.

Men stricken with disease also suffered badly. John Spencer Cooper wrote about his experiences suffering from fever and dysentery in 1809. He was taken to a convent where he was placed in a corridor with 200 others: 'My case was pitiable; my appetite and hearing gone; feet and legs like ice; three blisters on my back and my feet unhealed and undressed; my shirt sticking in the wound caused by the blisters; my knapsack and necessaries lost; and worst of all, no one to care a straw for me.' He was moved from hospital until he recovered, surviving on a diet of biscuit, salt pork and wine.

The commander of the British forces in the Peninsular War, the Duke of Wellington, realising that the unnecessary loss of men would affect the forces available to face the French, organised a new system, whereby more hospitals, and transport to get the wounded to them, were provided, and attempts were made to improve the care offered to the men.

The Battle of Waterloo, on 18 June 1815, was the bloodiest battle fought by the British Army before the First World War and showed medical deficiencies to the full. Foolishly the improvements in medical arrangements were quickly abandoned after the defeat of Napoleon in 1814, leaving Wellington to rue-fully regret that: 'the same reckless sacrifice of legs and arms has again taken place and nothing could undo the irretrievable mischief insufficient care has occasioned.'

The inadequacies were such that it was not until the following day that many of the wounded were rescued from the field of battle. In many regiments there were almost as many deaths as there had been on the 18th itself. In the 32 Regiment of Foot, which had lost twenty-eight men during the battle, another eighteen died on the 19th and between 27 June and 29 July another twenty-three were to die. The last direct casualty did not finally succumb to his wounds until 16 January 1816.

When the Army went to the Crimea, forty years later, the only medical and surgical aid provided was by the regiment. If the regimental surgeon could not cure the man, then he died. Within weeks it was clear that this was not satisfactory. The War Office tried to establish field hospitals and despatched 300 decrepit pensioners to establish a Hospital Convalescent Corps. Later on a Medical Staff Corps, comprising NCOs and private soldiers, was set up. These were supposed to be: 'Men able to read and write, of steady habits and good temper and of a kindly disposition,' but this level of perfection was rarely met.

That the appalling medical conditions in the Crimea were exposed was largely due to Florence Nightingale. She never went to the battlefields, but opened a hospital in the Constantinople (Istanbul) suburb of Scutari. The soldiers she cared for had to survive for days or weeks before they reached her. Because she was an outsider (and a woman to boot) her reforms, sensible though many of them were, were bitterly resisted by medical men in the Crimea. Miss Nightingale later grumbled that: 'their heads are so flattened between the boards of Army discipline that they remain old children all their lives'.

It was reports in *The Times*, submitted by William Howard and Thomas Chenery which finally alerted the public at home to the shambles on the Black Sea. Chenery wrote about conditions at the main British hospital at Scutari: 'Not only are there not sufficient surgeons – that it might be urged, was unavoidable – not only are there no dressers and nurses – that might be a defect of the system for which no one is to blame — but what will be said is that there is not even linen to make bandages for the wounded? But why could this clearly foreseen event have not been supplied?'

Things were made worse by the fact that the French allies had made better provision for the sick and wounded.

Medical treatment had advanced little since the Napoleonic wars. The young Garnet Wolseley, later a senior British Army commander, received shrapnel in the jaw after a shell burst close by. This was removed without anaesthetic: one doctor pulled it out with forceps while another held Wolseley's head between his knees.

Understandably the men had little faith in the official medical services. The West Indian nurse, Mary Seacole, set up a canteen in the British lines at Balaclava, which William Russell reported, attracted men 'who had faith in her proficiency in the healing art, which she justified by her many cures and by removing obstinate cases of diarrhoea, dysentery and other camp maladies'.

The authorities did take some action and gradually conditions improved, but they were hampered because the medical men and their orderlies were not under the direct command of the commander-in-chief.

One innovation was the prefabricated hospital designed by the great engineer Isambard Kingdom Brunel, which was kept cool by an ingenious ventilation system. It also contained kitchens, storerooms and ward rooms for the nurses.

After the war's conclusion a Royal Sanitary Commission was appointed, with Florence Nightingale as one of its members. It found that the mortality rate among soldiers was double that of the civilian population, with the Army at home losing 20 per cent of its strength each year to illness or disease.

Matters gradually improved at home and abroad. Barracks and hospitals were rebuilt or modernised and the use of anaesthetics and disinfectant became commonplace. Yet many of the old ways remained. Garnet Wolseley wrote in his memoirs of the doctors in Burma in the 1860s who: 'starved wounded men to keep down inflammation' and described a river boat on which doctors sat laughing, eating and drinking 'in the midst of their hungry and wounded'.

Medical men in the Army were generally of a low quality. They were poorly paid and often poorly treated, being expected to perform a variety of administrative duties and run the regimental officers' mess. It was not until 1898 that Army doctors were given Army ranks with the same titles as combatant soldiers. At the same time the Royal Army Medical Corps was established, which was an amalgamation of the Medical Officers (officers) and the Medical Staff Corps (other ranks).

The South African War (1899–1902) again revealed shortcomings in the British Army's medical arrangements. Here the threat was enteric and typhoid fever, caused by poor drinking water, rather than the more basic sanitary issues

A British square during the Sudan campaign of 1884.

that had been the cause of much illness in the Crimea and previous conflicts. Indeed Lord Robert's advance towards Pretoria in May and June 1900 came to a halt because of outbreaks of fever among his troops.

The problem was, in part, because of the resources available to treat the sick – there were no more than a couple of dozen lady nurses in South Africa at the height of the campaign in the early months of 1900 – but also because the inflexible and unimaginative way in which the Army Medical Service was run, owed more to red tape than to the Red Cross, as one historian put it.

Unfortunately for the Army the deficiencies made good copy. William Burdett-Coutts wrote in *The Times* of his experiences in Bloemfontein: 'hundreds of men to my knowledge were lying in the worst state of typhoid, with only a blanket and thin waterproof sheet (not even the latter for many of them) between their aching bodies and the hard ground, and no milk and hardly any medicines, without beds, stretchers or medicines, without linen of any kind, without a single nurse among them, with only a few ordinary private soldiers to act as "orderlies" and with only three doctors to attend on 350 patients'

What is less known is the disposal of the dead. Before the Crimean War there are almost no war cemeteries and few monuments to the fallen, although a number were erected at the battlefield at Waterloo. The dead and dying lay where they fell, until their bodies were dumped in mass graves. Only during the

Boer War was there any attempt to collect, identify and bury the dead in something approaching the cemeteries we are familiar with for the two world wars.

Scavengers would scour the battlefields searching bodies for anything which they could sell. Teeth torn from the dead at Waterloo, for example, were sold all over Europe and North America. The flood of teeth onto the market was so huge that dentures made from second hand teeth acquired a new name: Waterloo teeth. Far from putting clients off, this was a positive selling point. Better to have teeth from a relatively fit and healthy young man killed by cannonball or sabre than incisors plucked from the jaws of a disease-riddled corpse. A newspaper report of 1822 noted that amongst cargo landed at the port of Hull was a million bushels of bones recovered from the battlefields of Europe that were to be ground up and sold to farmers as fertiliser.

7.1 Casualty lists and rolls

Casualty lists were kept for many of the campaigns in which the Army fought during the second half of the nineteenth century. They include officers and men who died of disease (often the majority of deaths) as well as those who fell in battle.

A number of casualty lists have been published, often compiled from lists which originally appeared in the *London Gazette* or reproduced from original records at The National Archives. They include:

Frank and Andrea Cook, *The Casualty Roll for the Crimea* (London, 1976)
Ian T Tavender, *Casualty Roll for the Indian Mutiny 1857–59* (London, 1983)
— *Casualty Roll for the Zulu and Basuto Wars, South Africa, 1877–79* (London, 1985)
— *South Africa Field Force Casualty List, 1899–1902* (London, 1972), a facsimile of WO 108/338.

Copies of these books are available in the library at Kew.

Other published rolls include: MH Mawson, *Casualty Roll Egypt 1882* (Haywards Heath, 2000)

There are also several websites with (usually incomplete) casualty lists, such as www.roll-of-honour.com. Two worth visiting are http://members.tripod.com/~Glosters/obits1.htm (for officers, including some obituaries from newspapers) and the same webmaster's site for other ranks at www.angelfire.com/mp/memorials/memindz1.htm (sometimes including memorials and place of burial). There is also http://hometown.aol.co.uk/kevinasplin/homehtml, where the webmaster, Kevin Asplin, has usefully indexed a number of Victorian casualty rolls in surname order – the originals were normally arranged by regiment.

Regimental museums are beginning to post rolls of honour online, including the Worcestershires (for the 29 Foot) at www.worcestershireregiment.com.

Original casualty rolls, including some which have not been reprinted, are at TNA under the following references:

Campaign	Date	Reference
Burma	1888	WO 25/3473
China	1857–1858	WO 32/8221, 8224, 8227
China	1860	WO 32/8230, 8233–8234
China (Tsingtao)	1915	WO 32/4996B
Egypt	1882, 1884	WO 25/3473
New Zealand	1860	WO 32/8255
New Zealand	1863–1864	WO 32/8263-8268, 8271, 8276–8280
Sierra Leone	1898	WO 32/7630-7631
South Africa	1878–1881	WO 25/3474, WO 32/7700, 7706–7708, 7727, 7819
South Africa	1899–1902	WO 108/89-91, 338
Sudan	1884–1885	WO 25/3473, WO 32/6123, 6125–6126, 8382

7.2 Casualty returns

Series WO 25 contains several series of monthly and quarterly casualty returns for both officers and ordinary soldiers, usually arranged by regiment. There is often an index to surnames in each volume. These returns date between 1809 and 1910. They give name; rank; place of birth; trade; the date, place and nature of the casualty; debts and credits; as well as next of kin or legatee. Among these returns there may be copies of wills of deceased soldiers and inventories of their effects, correspondence with relations and statements of accounts.

In particular returns in WO 25/3250-3260 cover the period 1842–1872 and entry books of casualties (1797–1817) are in WO 25/1196-1358. These books, arranged by regiment, give the names in alphabetical order with details of cause of death and any financial credits the deceased might have had. These records have been microfilmed and are available in the Microfilm Reading Room at Kew.

7.3 Miscellaneous sources

Under the heading 'Deaths and Effects' (WO 25/2963-2978) there are various volumes listing the amount individuals (officers as well as men) left, together with details of the executors.

Register of effects for men killed during the Peninsular War. (TNA:PRO WO 25/2966)

A list of officers killed and wounded at the Battle of Waterloo is contained in Wellington's despatch of 29 June 1815, which was printed as a supplement to the *London Gazette* of 1 July 1815. Copies can be found in ZJ 1/138 and MINT 16/111.

Men who were struck off the regimental strength because of death or sickness are recorded in the muster rolls in WO 12–WO 16. Wounds and illnesses serious enough to lead to a man's discharge will be recorded on his soldier's document in WO 97: however none survive for men who died while in the Army. More details can be found in Chapter 5.

7.4 Hospital records

Before the 1850s, British Army medical services were organised on a regimental basis. Each regiment had its own medical officer and the male orderlies who staffed the regimental hospital were seconded from the regiment. The orderlies received no medical training but often acquired some knowledge through length of service.

During peacetime the problems of this decentralised local system were not apparent. It was well suited for dealing with minor skirmishes and policing duties in scattered locations throughout the world. There were only three military general hospitals at Chatham, Dublin and Cork. In addition the medical officers were part of the civil establishment of the Army and had no authority over the orderlies, who were classified as combatants.

The experience of the Crimean War (1854–1856) highlighted the difficulties of the peacetime organisation. The inadequacies of equipment and supplies, and the problems caused by poor communications, were compounded by the lack of medical experience of many of the orderlies, and by poor management of resources.

Few records of individual military hospitals survive. It is even rarer to find papers relating to individual patients. If you know a hospital in the UK where a man was treated, then it is worth checking the Wellcome Institute/TNA database of hospital records at www.nationalarchives.gov.uk/hospitalrecords to see whether anything survives.

One exception, however, is the muster rolls for the hospital at Scutari near Constantinople (Istanbul) in Turkey, where many sick and wounded soldiers from the Crimea were treated. They can be found in series WO 15 at Kew.

The largest (indeed the largest hospital ever built) was the Royal Victoria Hospital at Netley which stretched a quarter of a mile along Southampton Water. It was opened in 1866, in response to the appalling way in which the sick and wounded had been treated in the Crimea, and finally closed a century later. Its story is told in Philip Hoare's excellent *Spike Island: the Memory of a Military Hospital* (Fourth Estate, 2001). The hospital's records are largely with the Wellcome Library (of particular interest are the clinical and patients' records which survive for the period between 1866 and 1883), although The National Archives has casualty returns in WO 25/3260 between 1866 and 1871 and muster rolls between 1863 and 1878 in WO 25/13077-13105.

The Wellcome Library in London, also has the archives of the Royal Army Medical Corps and its predecessors. You should contact Archives and Manuscripts, Wellcome Library, 183 Euston Road, London, NW1 2BE. Tel: 020 7611 8888, www.wellcome.ac.uk

The Army Medical Services Museum also has an extensive archive and they may be able to help you find out more about ancestors who served in the Corps and tell you about developments in Army medicine, which may help you understand what happened to your ancestor. Contact Army Medical Services Museum, Keogh Barracks, Ash Vale, Aldershot, GU12 5RQ, Tel: 01252 340320, www.army.mod.uk/medical/ams_museum/index.htm.

Nurses' records are described in Chapter 11.

7.5 Memorials

Monuments to the fallen located in the British Isles have largely been recorded by the United Kingdom National Inventory of War Memorials. The results have now available online at www.ukniwm.org.uk. They have, for example, traced nearly 100 memorials for the Crimean War. Some memorials record the names of the fallen (particularly officers), but most are just for a regiment or commemorate the contribution from a town or village.

Records for the twentieth century are somewhat different (and rather more comprehensive) and you need to check the appropriate chapters for the two world wars.

Chapter 8

DISCIPLINE AND DESERTION

Foremost in the minds of the military authorities was discipline. The need for the ordinary soldier to follow the commands of his officers could be vital in battle. Soldiers grumbled about the discipline – the unnecessary drills and the pointless polishing of kit – but usually realised there was a point to it. In his autobiography Stephan Graham of the Scots Guards made clear that: 'for private soldiers in action the one thing needful is obedience. Imagination, thought, fear, love and even hate are out of place, and through stern discipline can be excluded. Discipline is the necessary hardening and making dependable . . .'

Throughout the Army, discipline and steadiness were considered vital; initiative and intelligence were looked upon as civilian qualities and therefore suspect. This could be a virtue – there are many accounts of battles being won as the square or the line maintained its cohesion against the odds. But the downside was that when original thought was required, perhaps with the loss of officers in the heat of battle (such as on Majuba Hill in South Africa in 1881), or in a situation for which the men had not been prepared, the Army did not perform as it should. JH Stoqueler, writing in 1857 summed this up when he said: 'the helplessness of the British soldier when left to himself is proverbial.'

In peacetime the generally poor quality of recruits and the brutalising conditions in which soldiers lived in the eighteenth and first half of the nineteenth century, meant that discipline was a constant problem. One former guardsman, Alexander Somerville wrote in his autobiography published in 1848 that 'nothing but the most severe punishment could have any influence on some of the ruffians that had enlisted solely for cheap spirits and were born mischief makers'.

The problem was fuelled by the ready availability of alcohol and the general lack of more sober alternatives. Drunkenness was the most common reason why a soldier was put on a charge. And drink was often behind the other common disciplinary problems of desertion, insubordination and petty theft.

It was perhaps little wonder that the better sort of pubs in garrison towns occasionally displayed notices 'no dogs or soldiers'.

Desertion was a major problem for the Army. Men unhappy at the conditions they had to endure or having decided that they no longer wished to serve the crown, would leave their posts. In many cases this was almost involuntary – many men deserted fearing the punishment when they found themselves overstaying furlough or even returning late from the pub one evening.

Desertion was particularly common in the months after a man enlisted. A study of the 38 Regiment of Foot (Infantry) in 1768 showed that of forty men who had enlisted during the year, fourteen had subsequently deserted. Some recruits made a habit of accepting the bounty money ('the King's shilling'), deserting and repeating the process with another regiment. Otherwise most men deserted, in the words of one writer, because of: 'Hardship and boredom; marauding and drink.'

Desertion on active service was always a problem. In the Peninsular War, Wellington lost roughly 500 men a year. Across the whole of the British Army in 1809, 14,476 men died in action or as the result of sickness, 2,968 were discharged as being no longer fit for service and 4,186 took matters in their own hands and deserted. Many deserted to the enemy in the hope of finding better conditions. At the siege of Badajoz, the French commander instructed his officers that British deserters in their units should wear their red uniforms, to ensure that they would fight well, because if they were captured they would expect to be executed.

Punishment could be severe for men who were caught. Until the 1870s it was the death penalty, but this was rarely applied. Often the man was flogged or on occasion transported overseas. More commonly the offender would be imprisoned or confined to barracks.

An alternative, until it was abolished in 1871, was to be tattooed with the letter D, according to the regulations: 'two inches below and one inch in rear of the Nipple of the Left Breast' and the needle punctures were to be impregnated 'with some ink or Gunpowder, or other preparation so as to be clearly seen'.

Until it was abolished in 1881, punishment for most disciplinary offences was generally by flogging. For minor offences the punishment was twenty-five lashes, but for more serious offences the sentence was left to the caprice of the commanding officer. Until 1807 when an Army Order laid down that the maximum punishment was 1,000 lashes, sentences of 1,500, 1,800 or even 2,000 lashes were quite common.

Following an inquiry in 1834 the maximum number of lashes which could be awarded by a general court martial was reduced to 200; a district court martial could award 150 and a regimental 100. Flogging could only be used to punish mutiny or insubordination, drunkenness on duty, and the theft of Army property or stealing from comrades.

There was a public outcry after Private Frederick White of the 7 Hussars, died after receiving 150 lashes at Hounslow in June 1846. The coroner's jury who investigated the death condemned 'the disgraceful practice of flogging' and the War Office had to announce that the maximum number of lashes which could be awarded by any court martial was henceforth limited to fifty.

Despite mounting public pressure flogging was not banned until 1881. In its place was introduced Field Punishment No. 1, whereby the offender was lashed to a gun- or carriage wheel for up to four hours a day in all weathers, while during his periods of release he was subjected to pack-drill and extra fatigues. As the name suggests this punishment was limited to active service. The maximum term an offender could be awarded was twenty-one days. It in turn was abolished in the late 1920s.

In order to create the maximum impression on the men, all floggings were performed at a parade of the whole of the victim's unit, which was formed up in a hollow square. In the centre stood the 'triangles' originally sergeants' halberds, but later a timber structure to which the prisoner was lashed, stripped to the waist. Also present was the commanding officer, the adjutant and the medical officer. The punishment was delivered by the regiment's drummers under the watchful eye of the drum major. In order to ensure they remained fresh the men were relieved after every twenty-five strokes. If the victim fainted or was judged to be unable to receive the full punishment, the whole performance would be repeated once he had recovered.

It is unlikely that flogging had much effect on discipline – if anything it contributed to the brutalising nature of a soldier's life. This was recognised by Dr Henry Marshall, Deputy Inspector of Army Hospitals, who wrote in his history of Army punishment of 1840 that: 'It is a notorious fact that when flogging was at its height it was counted no great disgrace, indeed it was sometimes made a boast of, and instances have occurred where to have suffered under the lash was reckoned qualification enough for becoming a good comrade.' Thirty years earlier in Gibraltar it was reported that there was a drummer who during the course of his fourteen years' service had received 25,000 lashes and was reported to be 'hearty and well, and in no way concerned'.

With the reduction, and eventual abolition, of flogging, alternatives had to be found to punish offenders. They included military prisons, the first of which opened in 1844. One of the largest was erected in Aldershot in 1855. When reconstructed in 1870 the main roof over the three storey block of cells was entirely made of glass, so it became known as 'the glasshouse' – soon adopted as the nickname for Army prisons in general.

Other punishments which could be meted out included confinement to a cell in the guardhouse; confinement to barracks (known as CB) for a period not longer than a month, extra drill in full marching order (not exceeding fourteen days). But surviving soldiers' documents suggest that other common punishments were the loss of pay or being reduced to the ranks.

For the most serious crimes, such as murder and mutiny, there was the death penalty, although it was often difficult to find a hangman or enough men to form a firing party. In 1856 a man in the 80 Regiment who killed a gunner was sentenced to be hanged. Even though a free discharge and £20 was offered, it was difficult to tempt a volunteer. When at last one was found, he was locked in a stable for safety. He later changed his mind.

Shooting parties were formed at random from within the offender's own unit. Usually between eight and twelve men were selected. The rifles would previously be loaded, one of which with a blank round, the idea being that right up to the point of firing each member of the firing party could continue to hope that he alone might be absolved from playing a part in the killing of a comrade.

As the nineteenth century progressed there was a marked decrease in capital punishment. In the decade between 1826 and 1835 there were seventy-six death sentences in the Army, of which thirty-three were commuted to transportation. In the period between 1865 and 1898, forty-four men were so sentenced and thirty-three executed.

An alternative was to be dismissed from the Army, usually in the most humiliating way. Neville Lyttleton described in his memoirs the manner in which men were drummed out in the 1860s:

> *The battalion was drawn up in two lines, five paces apart and facing inwards. The culprit was marched down between the lines, sometimes led by a drummer boy holding a rope round the man's neck, with band playing the 'rogue's march' behind him. On arrival at the barrack gate the boy kicked him out into the road where he was taken charge of by the police, his facings and buttons having been cut off. Occasionally he showed signs of shame and remorse, but by no means always. I heard of a culprit who, when leaving, bowed to his friends in the ranks as he passed them, and once a man was reported as saying to the Colonel 'A fine battalion, sir; you may dismiss them. I do not require them anymore.'*

As conditions improved, particularly with the introduction of short service, which meant it was possible for men to join up for a limited period thus encouraging better educated men to enlist, the number of punishments and their severity declined.

The *Journal of the Household Brigade* for 1874 looked at the reasons for desertion and found: 'of the 743 soldiers sentenced last year for desertion, 229 alleged they disliked the Army, 57 were annoyed by comrades or badly treated by non-commissioned officers, 44 married without leave or had "love affairs", 87 through drink, 39 on the "spree" and 18 through "whim and folly".'

In 1874 there were twenty desertions per thousand men (itself half of that of ten years previously); by 1893 it had declined to twelve per thousand. In 1869 there were 144 courts martial per thousand men; in 1892 it was fifty-four per thousand.

In addition, as we have seen, the amount and variety of punishment which regimental officers, and for the more serious offences courts martial, could administer were much reduced.

The Records

8.1 Courts martial

There were three different types of court martial: general courts martial, general regimental courts martial (before 1829) and district courts martial (after 1829).

The general court martial (GCM) was the Army's highest tribunal, dealing with commissioned officers and the most serious cases involving other ranks. It could only be convened by the crown or by overseas commanders or colonial governors. At least thirteen commissioned officers had to be present if 'at home' (that is in the British Isles, Ireland, non-British territories or small British possessions), or five if 'overseas' (the British colonies and India), together with a judge advocate. Decisions were confirmed by the person who issued the warrant (that is, the crown or its direct deputy).

Field general courts martial (FGCM) were often used in wartime. Only three commissioned officers needed to be present. The decision had to be unanimous for the death penalty to be imposed.

General regimental courts martial (GRCM), or district (or garrison) court martial (DCM), were more limited in jurisdiction. These courts could not try commissioned officers or charges carrying the death penalty, transportation, floggings of more than 150 lashes or prison sentences of more than two years. They were replaced by the district court martial, in 1829. It required seven officers at home or five if overseas. Details of the sentence were sent up to the Judge Advocate General's Office.

Regimental courts martial (RCM) were used for other ranks charged with lesser offences. No records were sent to the Judge Advocate General's Office. Some of the records of these courts may survive at regimental museums or archives.

Tracing the courts martial of commissioned officers is relatively straight-forward, since they could be tried only by general court martial. WO 93/1B is an index to trials of officers between 1806 and 1904. WO 93/1A is an index to general courts martial between 1806 and 1833.

There are three main types of record relating to individual trials: papers, proceedings and registers. Papers were compiled at the time of the court martial and are arranged in date order. They are in WO 71/121-343 (1688–1850), with one file for 1879. Other papers for trials between 1850 and 1914 were destroyed by enemy bombing in 1940. Papers for some special cases, mainly senior officers, are listed individually between 1780 and 1824 in WO 71/99-120, as are special returns for Ireland, 1800–1820, which are in WO 71/252-264.

When papers reached the Judge Advocate General's Office, their contents were entered into the volumes of proceedings. They were kept in two series depending on whether the sentence was confirmed at home by the Sovereign, or abroad by a colonial governor or overseas commander. These records are in WO 71/13-98 and continued in WO 91.

Until the mid-nineteenth century, the proceedings report the trials in detail, but later volumes give only the charges, findings and sentences in the form in which they were handed to the Sovereign. They also contain copies of warrants for the holding of courts martial and correspondence concerning the confirmation of sentences. Registers of warrants are in WO 28. The commander-in-chief's submissions upon sentence are in WO 209.

As well as volumes of proceedings, the Judge Advocate General's Office compiled registers of courts martial, giving the name, rank, regiment, place of trial, charge, finding and sentence. Registers of courts martial confirmed abroad are in WO 90 and those confirmed at home are in WO 92.

Records of field general courts martial date only from the South African War (1899–1902) and are combined in registers with district courts martial for 1900 and 1901, in WO 92. Later registers, between 1909 and 1963, are in WO 213.

NCOs and ordinary soldiers could be tried by general regimental courts martial (before 1829) and district courts martial (after 1829), as well as by general courts martial. As a result it is more difficult to find the records of individual cases. Only registers, rather than full proceedings, were compiled in the Judge Advocate General's Office. Registers of general regimental courts martial, between 1812 and 1829, are in WO 89 and of district courts martial, between 1829 and 1971, are in WO 86. Both classes contain trials confirmed both at home and abroad, except those for London for the ten years after 1865, which are in WO 87, and India, 1878–1945, in WO 88.

For general courts martial the records are as described for officers above. More information can be found in TNA's research guide *British Army: Courts Martial, 17th–20th Centuries*.

A Second World War court martial. (Taylor Library)

8.2 Deserters

There are numerous records about deserters at The National Archives. For the period before 1830 there is an incomplete card index compiled from bounty certificates of rewards paid out of locally-collected taxes to those who had turned in deserters. The index covers only rewards paid out in London, Middlesex,

Bedfordshire, Berkshire, Buckinghamshire, Cambridgeshire and Cheshire. The main part of the index is of deserters, giving date and regiment, as well as a reference to E 182 by piece number and sub-number. There is also a sequence by county, as well as cross-references from entries like 'Dragoons', 'Fencibles' and 'Militia' to the main sequence.

Registers of deserters between 1811 and 1852, are in WO 25/2906-2934. Until 1827 these volumes consist of separate series for cavalry, infantry and Militia (the latter up to 1820 only). After 1827 they are arranged in one series by regiment. They give descriptions, dates and places of enlistments and desertions, and may indicate what happened to deserters who were caught.

Registers of captured deserters, from 1813 to 1848 and indexed to 1833, are in WO 25/2935-2954. They include registers of deserters who were caught or who surrendered, and give the name of the individual and his regiment; the date of his committal and place of confinement; what happened to him (that is, whether he returned to his regiment or was discharged from the Army); and the amount of the reward paid (if the man had not surrendered) and to whom paid.

The police newspaper *Hue and Cry* (later renamed the *Police Gazette*) carried details of deserters, giving name, parish and county of birth, regiment, date and place of desertion, a physical description and other relevant information. The National Archives has copies of *Hue and Cry* between 1828 and 1845 in HO 75 and a selection of *Police Gazettes* from the 1834 to 1959 in MEPO 6. The British Newspaper Library at Colindale has editions of *Hue and Cry* going back to 1801 as well as a set of the *Police Gazette* and it may be possible to find sets elsewhere.

Men who deserted the Army in Australia are listed in Yvonne Fitzmaurice, *Army Deserters from HM Service* (Forest Hill, Victoria, 1988). Details of men who deserted from the British Army in Canada during the War of 1812 and settled in the United States can be found at http://freepages.history.rootsweb.com/~british41st/41stregt_deserters.htm. There is also a chapter about deserters in Ruth Paley and Simon Fowler's *Family Skeletons* (TNA, 2005).

Chapter 9

PENSION RECORDS

For many men leaving the Army was a traumatic experience, particularly for those who served for many years. Where once there had been discipline and the certainty of the regimental family, there was now uncertainty and the difficulties of dealing with undisciplined civilians whether members of the soldier's own family or new work colleagues. It is little wonder that so many unmarried and unsettled veterans ended up as tramps.

Matters could be worse for veterans who left the Army with a serious disability, for there was little support until the 1880s when charities such as the Soldiers' and Sailors' Help Society and the Lord Roberts' Workrooms were established to help the disabled serviceman.

The first such charity was the Corps of Commissioners set up in 1859, by Edward Walter, to help wounded ex-soldiers, who were found posts as hotel doorkeepers, night watchmen and bank messengers. By 1907 when it was inspected by Edward VII at Buckingham Palace there were more than 3,000 men on its books. It is, of course, still in existence.

From the 1680s until the First World War looking after veterans was largely the responsibility of the Royal Hospital Chelsea. When Chelsea was founded in 1681 (its Irish equivalent at Kilmainham outside Dublin was founded in 1679) the intention was that it would house all old soldiers. In practice this was never possible, so there was soon a division between in-pensioners – the men who lived at the hospitals – and out-pensioners. In 1888, for example, there were 87,703 out-pensioners and just 550 in-pensioners at Chelsea.

The pension was originally raised from a levy of one day's pay a year on soldiers and a tax on the commissions bought and sold by officers. However from 1731 this was largely replaced by a parliamentary grant. This enabled 5d a day to be paid to the few men who survived Army service to receive a pension.

The Royal Hospital Chelsea which was responsible for Army pensions until the First World War.

Corporals received 7d and sergeants 11d a day. By the 1830s this had risen to 1s per day for privates with an extra 1½d for corporals, provided he had completed twenty-one years' service or twenty-four in the cavalry.

Within a few years the payment of a guaranteed pension had become a considerable attraction in recruiting soldiers. It was considered, said Charles Grose in 1801: 'particularly in the North as a comfortable provision for old age.'

In time of national emergency pensioners could be required to guard forts and other military establishments. This was not a great success as most men were so 'worn out' (as the phrase went) that they were almost worse than useless. If they were unable then they stood the risk of losing their pension. In 1815, Rifleman Benjamin Harris, who had suffered from malaria as the result of service in Holland and, as a consequence, had been recently discharged said: 'However, before my pension became due and in consequence of the escape of Bonaparte from Elba, I was with others called upon to attend. But I was in so miserable plight with the remains of the fever and ague – which still attacked me every other day – that I did not answer the call and thereby lost my pension.' Recent research has found that Harris's pension was eventually restored in 1854, four years before his death.

For men who left the service disabled, the King's 'Bounty to Soldiers' for the loss of an eye or a limb or the total loss in the use of a limb was, in 1685, one year's pay with 'other wounds in proportion'.

By the 1830s if the veteran was so severely wounded (that is by the loss of two limbs or both eyes) that he was unable to support himself, he would receive a pension between 1s 6d and 2s 0d a day, although the exact amount depended on how many years he had served.

Further changes were made in the 1890s. The private soldier who had served continuously for twenty-one years received 1s 1d per day, although this was reduced to 10d if he had only put in eighteen years' service. In practice most men would have been better off as they received extra for having specialist skills and the award of good conduct medals.

Pension Records

Originally pensions took the form of accommodation for disabled soldiers in the Royal Hospital Chelsea or the Irish establishment at the Royal Hospital Kilmainham (now home to the Irish Museum of Modern Art). These pensions became known as in-pensions.

Within a few months the accommodation became insufficient to meet the demand, and a system of out-pensions for non-residents was devised to supplement the original in-pensions. They could be claimed on the grounds of disability or unfitness arising from service. In the 1750s regulations were passed to make length of service and character the principal reasons for award of pensions, and not disability. Responsibility for out-pensions of Irish pensioners passed to Chelsea in 1922, and for in-pensioners in 1929.

Except for a few officers admitted as in-pensioners, the two hospitals were not concerned with officers' pensions. Chelsea's role in paying out-pensioners ceased during the First World War with the establishment of the Ministry of Pensions, although of course it remains a home for old soldiers.

The history of Chelsea Hospital is described in Dan Cruickshank's *The Royal Hospital Chelsea: the Place and the People* (London: Third Millennium, 2004) and David Ascoli's *A Village in Chelsea: an Informal Account of the Royal Hospital* (London: William Luscombe, 1974). There are no similar accounts for Kilmainham Hospital.

More about pension records can be found in the TNA Research Guide *British Army: Soldiers' Pensions, (1702–1913)*.

9.1 Out-pensions

There are three main series of records containing information about out-pensioners: admission books, regimental registers and pension returns. These series cover the vast majority of pensioners at home and abroad.

Application for an out-pension from Royal Hospital Chelsea for William Perry, late 3 Foot, 1790. (TNA:PRO WO 121/140)

9.1.1 Admission books

Admission books are arranged chronologically according to the date at which the pension board met, which was generally the date a man was discharged from the Army. Therefore you need to know at least the approximate date of the award of a pension before a search becomes practicable. In each case there is a brief description of the pensioner, with his name, rank (if other than a private), regiment and observations, such as 'cured', 'struck off', or 'D' for discharged. The most common entry was 'DD' for discharged dead. Occasionally the date of death is given as well.

For pensions awarded by the Royal Hospital Chelsea, admission books covering pensions awarded for disability between 1715 and 1913 are in WO 116/1-124, 186–251. Details of pensions awarded for length of service in the Army, between 1823 and 1913, are in WO 117.

Records about pensions awarded by the Royal Hospital Kilmainham between 1704 and 1922 are in WO 118, with certificates of service (similar to Soldiers' Documents) to 1822 in series WO 119. Kilmainham looked after soldiers who lived in Ireland after they left the Army. There were about 250 in-pensioners and several tens of thousands of out-pensioners.

Between 1830 and 1844 the Chelsea admission books are duplicated by registers in WO 23/1-16 which also give the intended place of residence. The registers between 1838 and 1844 (in WO 23/10-16) are indexed.

9.1.2 Regimental registers

These registers are for pensions issued at Chelsea in two distinct series in WO 120. They are now available only on microfilm.

The first series (1715–1843) in WO 120/1-51 is arranged chronologically within regiments, and gives date of admission, age, a brief record of service, rate of pension, 'complaint', place of birth and a physical description. The volumes for 1839 to 1843 are indexed. In addition, there is a name index to some infantry regiments, between 1806 and 1858, in WO 120/23-26, 29–30.

A second series, in WO 120/52-70, records pensions being paid between 1814 and 1857. Admissions before 1845 are arranged by rate of pension, those between 1845 and 1857 chronologically. The registers give rate of pension, date of admission and residence, and note the place of payment of the pension and date of death. Later registers to 1876 are in WO 23/26-65.

9.1.3 Pension returns

Before 1842 out-pensions were generally paid by tax collectors in Great Britain and postmasters elsewhere. In 1842 payment was transferred to pensioners' staff officers. Each staff officer now made a monthly return to the War Office in which he recorded pensioners who had moved into, or out of, his district, whose pension had ceased, or who had died. Pension returns in WO 22 record pensions paid or payable from district offices between 1842 and 1883.

These returns give the pensioner's name, regiment, rate of pension, date of admission to pension, rank, and the district to which, or from which, he had moved. Also included with the returns are various items of statistical information. Returns for British payment districts cease in 1862, but returns relating to pensions paid overseas and in the colonies extend into the 1880s.

9.2 In-pensions

There are muster rolls for in-pensioners of Chelsea Hospital (the men commonly called Chelsea Pensioners) between 1702 and 1789 in WO 23/124-131, and for 1864 and 1865 in WO 23/132. A list of in-pensioners between 1794 and 1816 is in WO 23/134, with an alphabetical register of inmates (1837–1872) in WO 23/146.

Admission books for the periods 1778–1796 and 1824–1917 are in WO 23/133, 163–172, 174–180. Arranged chronologically, these books give regiment, name, age, service, rate of pension, cause of discharge, date of admission to pension, and decision of the Board of Chelsea Hospital. In addition, an address is often given. An index of in-pensioners admitted between 1858 and 1933 is in WO 23/173.

A list of in-pensioners at the Royal Hospital Kilmainham between 1839 and 1922 is in WO 118/47-48. This list also includes out-pensioners.

9.3 Other sources

Additional information about both in- and out-pensioners at Chelsea can sometimes be found in the Board Minutes and Papers (1784–1953) in WO 250 and the Invaliding Board Minutes and Papers (1800–1915) in WO 180, especially where appeals were made against decisions on eligibility for a pension and the rate at which it was to be paid. Incomplete minute books in WO 180/53-76 include appeals as well as other relevant papers.

Series WO 131 includes details of men who left the Army, but received their pensions sometime after discharge. They are arranged in name order and cover the period from 1838 to 1896. There are also some nominal rolls in chronological order.

Some 5,000 personal files of soldiers (and sailors) who received a disability pension and who left the armed services before 1914 are in PIN 71. A number of papers for widows' pensions are also included in this class. The files contain medical records, accounts of how and where illness or injuries occurred, and men's own accounts of incidents in which they were involved. Conduct sheets are included, recording the place of birth, age, names of parents and family, religion, physical attributes and marital status.

Pensions were also paid to former British soldiers who had emigrated to the colonies which are recorded in Chelsea Hospital out-pension registers (1814–1857) in WO 120/69-70.

Lists of men who emigrated to Australia and New Zealand between 1830 and 1848 under schemes to settle soldiers there are in WO 43/542 (for Australia) and WO 43/853 (New Zealand). Land grants made to soldiers, particularly officers in Canada in 1837 and 1838 are in CO 384/51.

In particular, areas around Auckland were settled by 750 former soldiers who formed the Royal New Zealand Fencibles who arrived in between 1848 and 1853. There is a website devoted to these men at http://homepages.ihug.co.nz/~rhaslip/Lesley/fencibles.htm. Archives New Zealand in Wellington also has a small collection of records which are described at http://www.archives.govt.nz/docs/pdfs/Ref_Guide_Migration.pdf.

Chapter 10

MILITIA 1757–1914

The Militia, under various names and forms, has existed since Saxon times. Initially every able-bodied man was expected to serve in times of national danger, although the effectiveness of this was never tested (see Chapter 3). Increasingly, the Militia was made up of men who either wanted to soldier part time or, during the nineteenth century, were serving in the reserves after time in the Regular Army.

The modern Militia originated with the Militia Act 1757 which established one or more regiments for each county, raised from volunteers and conscripts chosen by ballot from each parish. Officers were appointed by the Lord Lieutenant of the county. In peacetime the Militia assembled for drill and manoeuvres at intervals. After 1782 they came under the ultimate authority of the Home Secretary. In wartime however, having been mobilised (or embodied) by royal proclamation, they were subject to the orders of the commander in chief and were liable to serve anywhere in the British Isles but not overseas.

The Fencible Infantry and Cavalry, which were regular regiments raised for home service during the Napoleonic Wars, are often classed with the Militia.

Between 1804 and 1813 supplementary Militia and volunteer units were raised. The men who served in them were not conscripts but in other respects were similar to the Militia. In addition, however, they included cavalry units (the Yeomanry) and a proportion of artillery.

In the century thereafter there grew up a variety different Militia forces. At the time of the Boer War, the total strength of the regular Army based in the United Kingdom was 130,000 to which, on call-up, could be added 78,000 regular reservists (that is men who, although having left the Army, could be recalled in time of national emergency) and 30,000 members of the Militia Reserve. The 65,000 strong Militia could be called upon if they volunteered, but the 10,000 Yeomanry and 230,000 volunteers could not legally be sent abroad.

The oldest and most prestigious of these bodies was the Militia. During the Napoleonic Wars all adult men were liable for conscription into it, but this requirement faded away after Waterloo. By the middle of the century it was entirely made up of volunteers, who enlisted for six year periods, and carried out twenty-eight days' annual training with the regiment. They received an annual bounty of £1 as well as basic pay.

For an extra bounty of £1 they could join the Militia Reserve. It became a popular method of entry into the Army itself, as if the recruit found that he did not like Army life he could resign from it much more easily (and rather cheaper) than having to buy himself out of the regulars.

Soldiers in the Militia Reserve were usually either very young or very old, and were generally unskilled labourers. Officers were on the old side and came from a wide range of middle and upper class backgrounds.

Until the arrival of effective policing across Britain the Yeomanry was generally used in controlling unrest and disorder. Their least glorious day was at Peterloo in Manchester on 16 August 1819 which saw fourteen men and women killed and hundreds wounded as they swept through a peaceful demonstration of 60,000 calling for parliamentary reform.

By the end of the nineteenth century Yeomanry units regarded themselves as being the elite. Certainly they were the most socially prestigious. Members generally came from landowning and farming circles.

This was certainly not the case with the Manchester and Salford Cavalry Yeomanry, who led the charge at Peterloo. The unit had been formed in 1817, and was commanded by a retired officer Major Thomas Trafford. Hugh Birley, a factory owner from Manchester, was given the rank of captain and became Major Trafford's second in command. The occupations of the men of other ranks included shopkeepers, publicans, watchmakers, insurance agents, tobacconists, farriers, horse-breakers and brewers. However, a historian RJ White later observed that: 'The Manchester and Salford Yeomanry consisted almost exclusively of cheesemongers, ironmongers and newly enriched manufacturers, and the people of Manchester and district thought them a joke, and not a very good joke.'

In 1859, as a result of local pressure and fears of a possible foreign invasion, volunteer regiments were again formed. They were very much local formations of dubious efficiency, doing little else than basic drill and rifle shooting, to which they were generally addicted. In Brighton, staff of a department store joined one company of the Sussex Volunteer Artillery with the shop's owner and his family becoming officers. Another consisted of recruits from local friendly societies.

Although immensely popular with the middle classes they had little formal connection with the War Office (who regarded them as a nuisance) until 1873.

Thereafter, a small number of soldiers and officers or retired soldiers formed a permanent staff for each regiment.

As a result they generally raised much of their uniforms and equipment from their own pockets, although they received a grant for every man present at the annual parade. As late as 1911 the annual report of the 10 London Regiment (Paddington Rifles) noted that: 'A local recruiting committee of gentlemen living in Paddington raised round about £300', £50 of which was used to improve the other ranks' recreation room, and another £50 on 'other recruiting attractions'.

After the reorganisation of the Army on a territorial basis in 1881, the county Militia regiments became the third battalions, and the volunteer units the fourth and sometimes the fifth battalions of their local regiments. In 1908 the Militia was renamed the Special Reserve. The Volunteers and the Yeomanry (which later became the Territorial Force) were more closely integrated with it.

Membership of the Militia was a very social affair, particularly for the officers who came from a narrow stratum of the middle classes. Ordinary soldiers joined because it provided a way into the Regular Army or perhaps it was a way of having a paid holiday once a year. The fact that they were paid (if meagrely) may also have been a contributory factor.

By the early twentieth century each man joined for four years of spare-time training, which meant doing ten to twenty drills a year, plus two weeks' of training in camp and a musketry course. If called up in wartime they could choose whether they served just in the United Kingdom or overseas as well.

Except in times of national emergency numbers remained small, which always concerned the authorities. In July 1914 the Territorial Force numbered just under 269,000 officers and men, of whom only 19,000 had indicated a willingness to serve overseas in case of war.

The highlight of the year was the annual encampment, where units trained and paraded, often before appreciative crowds. Thousands of men attended mock battles and the like on the downs above Brighton, for example, in the 1860s. Here the highlight was the race to manoeuvre heavy artillery pieces up Race Hill aided and abetted by local people. Another popular feature was the shooting competitions. Both *Punch* and the *Illustrated London News* often carried stories about events at Bisley.

The Records

10.1 Officers
From 1865 officers appear in the official *Army Lists*. They were often retired Army officers on half-pay.

Information about appointment of officers to Militia and volunteer units from 1782 to 1840 can be found in the Home Office Military Papers (HO 50), with related entry books in HO 51. The papers include some establishment and succession books, but there is no name index and in general no information is given beyond the names of officers and the dates of their commissions.

Records of service of officers in a number of Militia regiments are in WO 68. They date from about 1757 to 1925, but are incomplete. Provided that the unit is known, it is possible to get a rough idea of an officer's service from the muster books in WO 13. Registers of pensions paid to Militia officers (1868–1892) are in WO 23/89–92. A selection of birth and baptismal certificates from 1788 to 1886 is in WO 32/8906-8913.

10.2 Other Ranks

The most useful records for the family historian are the attestation forms in WO 96. They range in date from 1806 to 1915, but the majority are from the second half of the nineteenth century. They are arranged alphabetically by surname order under the name of the regular regiment to which the Militia unit was attached. In form and content they are similar to soldiers' documents, which are described in Chapter 5. Many men went on to join regular Army units so you may also want to check the appropriate records in WO 97 and elsewhere.

Muster books and pay lists of the English, Scottish, Irish and Colonial Militia, and the Fencible Infantry and Cavalry, Yeomanry, Irish Yeomanry and Volunteers from 1780 to 1878 are in WO 13. Muster books and pay lists provide a means of establishing the dates of enlistment and discharge or death. When an individual appears for the first time the entry in the muster book may show his age. For volunteer units only, payments to the professional cadre, and not the ordinary volunteers, are included. Muster books are of use only if you know which unit your ancestor belonged to.

Records of payments to the families of those men who served in Militia units during the Napoleonic Wars are in series E 182.

A number of soldiers' documents for men who served in Militia regiments between 1760 (mainly from 1792) and 1854 are in WO 97/1091-1112. They are indexed in TNA's online catalogue.

10.3 Further information

The vast majority of other Militia records at Kew are in WO 68. They include order books, succession books, records of officers' service and enrolment books.

A number of muster books for units based in London and Middlesex are in WO 70. Records for a few provincial units are in WO 79. Records of the Tower Hamlets Militia are in WO 94.

Militia records at TNA are explained in more detail in William Spencer's *Records of the Militia and Volunteer Forces 1757–1945* (Public Record Office, 1997).

Some records of Militia and volunteer units may be preserved at local record offices, such as ballot forms for the Napoleonic Militia. The ballot paper for Rifleman Harris, for example, survives at the Dorset History Centre in Dorchester (Blandford Forum Militia papers ref PE/BF/OV8/1). On it he is described as being a cordwainer (shoemaker) who enrolled on 13 August 1803 for a bounty of £11. He signed for it with his mark, showing his illiteracy. The famous memoir of his adventures in the Peninsular War was in fact taken down by Henry Curling in 1848. Harris, himself, remembered that 'without troubling myself much about the change which was to take place in the hitherto quiet routine of my days, I was drafted in the 66th Regiment of Foot and bid goodbye to my shepherd companions.'

Jeremy Gibson and Mervyn Medlycott's *Militia Lists and Musters, 1757–1876* (4th edition, FFHS, 2000) lists the key sources to be found at local record offices. It is also worth trawling the Access to Archives database at www. nationalarchives.gov.uk/a2a to see what has been catalogued.

Regimental museums may well also have records for units with which they were affiliated.

If you are investigating Militia, volunteer or territorial units in any great detail it is well worth looking at the *Volunteer Service Gazette and Military Dispatch*, later renamed *Territorial Service Gazette* which was published between 1859 and 1925, and which contains a lot about the Militia and the men (especially officers). A set is held at the British Newspaper Library at Colindale and occasional copies may be found elsewhere.

Chapter 11

WOMEN AND THE ARMY

U ntil 1992, when women were finally admitted on full equality, the Army has always had an ambiguous relationship with the 'fair sex'. Women's involvement in the service formally began during the Crimean War, when a few were allowed to nurse sick and wounded men. During the two world wars increasing numbers of women were taken on initially to allow men to be sent to the front, but in the Second World War they undertook an important variety of duties from driving and cooking to acting as clerks.

11.1 Army wives

During the eighteenth and first half of the nineteenth centuries the Army in the field relied considerably on the unpaid and unacknowledged support of the wives of soldiers to cook and care for their menfolk. Six wives of soldiers in each company were carried on strength to act as unofficial cooks, laundresses and servants to officers. In addition it is clear that other women – wives and lovers – could also be present, as well as babes in arms and children.

Where women were permitted on 'the strength' wives could live inside the barracks, have half-rations and send their children to regimental schools. They often earned money by doing the regimental washing (for a halfpenny a day), preparing meals and where necessary an elementary nursing service. But the general lack of separate married quarters (only twenty out of 251 barracks in 1857) meant that marriages had to be consummated and babies born in the corners of barrack rooms screened by flimsy curtains.

The women shared the hardships endured by their men. The strength and endurance of some of these wives seems unbelievable today. During the Peninsular War for example, Dan Skiddy of the Rifle Brigade collapsed exhausted on the retreat to Vigo in northern Spain. His wife, Biddy, dragged him to a bank and struggled to get him on to her shoulders. Then she carried him, with his rifle and knapsack, two miles to the regimental base.

The last campaign in which (a few) women joined their men on campaign was the Crimean War. One of the women who made the journey was Margaret Kirwan, whose husband John was a private in the Green Howards. In the 1890s, she remembered that:

> *When we were up at Monastir [in Bulgaria] all the duty of No 5 Company fell upon me, that is the washing. There were 101 men in the Company and the clothes were brought down to the river by the transport horse. I stood in the midst of the stream washing the clothes from six in the morning until six or seven at night. The Colour Sergeant would keep no accounts and some men were able to pay and some were not, so that I was left with very little money for my trouble.*

For more about her story see, Roger Chapman, 'Margaret Kirwan – Woman of the Regiment' in *Soldiers of the Queen* (114) September 2003.

There are very few records of these Army wives. Occasionally women retained on strength may appear in the muster rolls in WO 12. Wives of soldiers are recorded in the soldiers' discharge documents, in WO 97, from the 1850s onwards.

11.2 Nurses

The first female nurses, famously recruited during the Crimean War (1854–1856), highlighted the difficulties of the peacetime organisation. The call for female nurses was popularised by *The Times*' war correspondent,

'The Lady with the Lamp': Florence Nightingale in the wards at Scutari in 1855.

WH Russell. Florence Nightingale was offered the command of a scheme to send a nurses' expedition to Scutari. The various religious sisterhoods were also keen to be involved. The first party sent out was made up of representatives from the Anglican and Catholic orders and secular nurses. Testimonials for nurses who wished to serve with Florence Nightingale during the Crimean War are in WO 25/264.

Despite the mixed reaction to female nurses in the Crimea, the draft Regulations for Inspectors General of Hospitals, 1857, contained a section on a code for female nurses to be employed in general hospitals. In 1861 Jane Shaw Stewart was appointed the first superintendent of a female nursing service, with a staff of six nurses, at Woolwich. Two years later she was also appointed Superintendent General of Female Nurses at the new hospital at Netley, thus becoming the first woman to appear in the British *Army List*.

After her resignation in 1868, the female nursing service continued, under Mrs Deeble, the widow of an Army medical officer. She and six nurses from Netley were employed in South Africa during the campaign against the Zulus in 1879.

The next military campaign which saw the use of female military nurses was in Egypt and Sudan where, between 1882 and 1885, approximately thirty-five women saw service. The nurses were recruited from many different hospitals with no one female superintendent in charge.

An Army Nursing Service was formed in 1881 and efforts were made to increase the numbers of female nurses in the Army. In 1883 the Army Hospital Services Committee recommended an expansion of the nursing corps. The service was to be extended to all hospitals with over 100 beds, female nurses were to help train the orderlies and, in wartime, a lady superintendent was to be appointed under the command of the Principal Medical Officer.

An Army Nursing Reserve was established in 1897. Both the Reserve and the Service were reorganised after the South African War (1899–1902) as Queen

A nurse helps a wounded soldier write home.

Alexandra's Imperial Military Nursing Service (QAIMNS). The present title of Queen Alexandra's Royal Army Nursing Corps (QARANC) was assumed in 1949.

11.2.1 Service records

Service records and correspondence for some of the women who served in QAIMNS and the Territorial Force Nursing Service during the First World War are in series WO 399. Records of nurses who served with the QAIMNS may contain details of service, enrolment and discharge papers, and correspondence relating to the period of service. The list gives forename and surname.

In common with the rest of the Army, service records for the Second World War are with the Ministry of Defence, Army Personnel Centre, Historic Disclosures, Mailpoint 400, Kentigern House, 65 Brown Street, Glasgow, G2 8EX, Tel: 0141-224-2023, email: apc_historical_disclosures@btconnect.com.

Nominal and seniority rolls for nurses in the voluntary National Aid Society and the Army Nursing Service, 1869–1891, are in WO 25/3955. An indexed register of candidates for appointment as staff nurses, 1903–1926, is in WO 25/3956.

Pension records prior to 1905 are in WO 23/93-95, 181. Few nurses qualified for a pension, however, because they rarely served enough years to receive one. Registers of pensions for nurses, 1909–1928, are in PMG 34/1-5. Awards for disability pensions are in PMG 42.

11.2.2 Medals

In 1883 Queen Victoria instituted the Royal Red Cross to be awarded to military nurses. Registers of the people to which it was awarded between 1883 and 1994 are in WO 145.

The Queen's and King's South Africa Medals were awarded to nurses for service during the South African War, 1899–1902; medal rolls are in WO 100.

Awards of campaigns medals to nurses during the First World War are in the medal rolls in WO 329. There is a separate name index to nurses on microfiche in the Microfilm Reading Room at Kew.

Nurse Elizabeth Mabel, from Angus, who served in hospitals during the South African War.

11.2.3 VADS

During the two world wars, the British Red Cross and St John's Ambulance supplied many volunteers, both male and female, to help in hospitals. The best known was Vera Brittain who famously wrote about her experiences in *Testament of Youth*. During the First World War they were generally referred to as VADS; in the Second World War as the Civil Nursing Reserve.

Record cards for members during both world wars are held by the British Red Cross Museum and Archives, 44 Moorfields, London, EC2Y 9AL, www. redcross.org.uk. The Archives has an informative online leaflet about VADs and how you can trace them. There is an article about VADs in the April 2006 (no. 44) issue of *Ancestors Magazine*.

11.2.4 Further information

Further information is contained in a TNA Research Guide: *British Army: Nurses and Nursing Services*.

The Army Medical Services Museum, Keogh Barracks, Ash Vale, Aldershot, GU12 5RQ, Tel: 01252 868612, www.army.mod.uk/medical/ams_museum has some records relating to QAIMNS and the women who served with it.

11.3 Auxiliaries

The Women's Army Auxiliary Corps (WAAC) was established in March 1917 to undertake clerical and manual work so releasing men for the front. In appreciation of its good service it was renamed the Queen Mary's Auxiliary Army Corps in April 1918. At its height in November 1918 the strength of the WAAC was over 40,000. Altogether nearly 57,000 women served in the Corps.

Unfortunately only about 9,000 service records survive. They are found in series WO 398 (on microfilm at Kew) arranged by surname. Their contents are similar to the service records for men. Officer equivalents were called 'officials', non-commissioned officers 'forewomen', and the rank and file 'workers'. These records are also available on TNA's Documents Online service. Visit www.nationalarchives.gov.uk/documentsonline for more details.

An incomplete nominal roll for members of the Corps is in WO 162/16 with a list of women drivers employed during the war in WO 162/62. Recommendations for honours and awards are in WO 162/65.

The Women's Service Medal lists give the names of members who were entitled to the various campaign medals, particularly the British War and Victory Medals. Again the information given is similar to the male equivalents, with service number, theatres of operation and perhaps a note about when the medal was issued.

During the Second World War, many young women, including Princess Elizabeth, joined the Women's Auxiliary Territorial Service (ATS). Service records are still with the Ministry of Defence (for address see above). Few records about the ATS are to be found at The National Archives, but there may well be material at the Imperial War Museum and the National Army Museum. An excellent site on the ATS is at www.atsremembered.pwp.blueyonder.co.uk.

Further reading
Lynn Macdonald's *Roses of No Man's Land* (Papermac, 1990)
Juliet Piggott, *Queen Alexandra's Royal Army Nursing Corps* (Leo Cooper, 1990)
Roy Terry, *Women in Khaki: the Story of the British Women* Soldier (Columbus Books, 1988)
Annabel Venning, *Following the Drum: The Lives of Army Wives and Daughters* (Headline, 2005)

Chapter 12

THE BRITISH IN INDIA

The British Army had a long connection with India, particularly over the ninety years between the Indian Mutiny in 1857 and the grant of independence to India and Pakistan in August 1947, when battalions were stationed across the sub-continent.

In addition there was the British officered Indian Army, which was under the control of the Viceroy in Calcutta or New Delhi. It was established in 1859 as the successor to the Army of the Honourable East India Company (HEIC), a private company which had grown to dominate India and Indian trade. It was defined by Lord Kitchener, who reorganised it in the late 1880s as being: 'the force recruited locally and permanently based in India, together with its expatriate British officers.'

It is easy confuse the two and assume that a British soldier serving in India was with the British Army. This is the case if you are researching an ordinary soldier or non-commissioned officer of European stock who served in the sub-continent after 1857. However officers could have been in either Army. It is easy to tell from the name of the regiment in which he served or by checking the *Army Lists.*

12.1 East India Company and Indian Army
From the mid-eighteenth century, the East India Company began to maintain armies at each of its three main stations, or Presidencies, in India at Calcutta (Bengal), Madras and Bombay. The Bengal, Madras, and Bombay Armies were quite distinct, each with its own list of regiments and cadre of European officers. All three armies contained both European regiments, in which both the officers and men were Europeans, and a larger number of 'Native' regiments in which the officers were Europeans and the rank and file were Indians.

From the mid-eighteenth century the British Government sent regiments of the regular British Army to India to reinforce the company's armies. These troops are often referred to as 'HM's regiments' or 'Royal regiments'.

Following the Indian Mutiny of 1857, and the consequent abolition of the East India Company, its European regiments were amalgamated in 1860 with the British Army. However, the Native regiments were not. The separate Presidency Armies therefore continued to exist, and their European officers continued to be listed as members of the Bengal, Madras or the Bombay Army rather than the British Army. However, the Presidency Armies began to be described collectively as the Indian Army. Another change resulting from the Mutiny was that artillery was confined to the British Army.

In 1889, the separate Presidency Armies were at last abolished and a fully unified Indian Army came into being. Lists of regiments in the Indian Army in 1903 and 1922 can be found on Wikipedia (http://en.wikipedia.org). Wikipedia also includes histories of a number of Indian Army regiments.

In the years before the First World War life for officers of the Indian Army was generally very pleasant. Apart from occasional campaigns against recalcitrant tribesmen on the Khyber Pass, the only activity they took seriously was sport — generally polo or big game hunting. It was, of course, argued that sports such as pig-sticking or boar hunting were good preparation for war. Major Robert Baden-Powell enthused that 'pig-sticking is nevertheless par excellence a soldiers' sport; it tests, develops and stands without rival as a training ground for officers.'

A Zhob Field Force officer in 1898.

Not everyone benefited. Colonel William Gatacre, in the 1890s, was an enthusiastic member of the Bombay Jackal Club. He was once bitten by a jackal, becoming deranged. Consequently he had his bungalow windows barred to prevent jackals jumping in and frightening him. It did him no permanent harm for he became a senior commander during the South African War.

As well as service in the sub-continent, the Indian Army gave good service during the two world wars. During the Great War it served first on the Western Front (where wounded Indian troops were cared for in the Brighton Pavilion) and then was transferred to fronts in the Middle East at Gallipoli and in Mesopotamia. In the Second World War the Indian Army became the largest all-volunteer force in history, rising to over 2.5 million men in size, fighting in Burma, Eritrea, the Middle East and Italy.

With the approach of independence the officer corps had an increasing number of Indian officers, although none reached a senior rank before 1947. The first staff college to train Indian officers was established in 1932.

The bravery and commitment of the troops to the Raj is shown in the fact that 143 Victoria Crosses were won by members of the Indian Army between 1857 and 1947.

The records

12.1.1 Officers

It is important to remember officers were not members of the British Army, although as young subalterns they did serve for a year with a British Army regiment as part of their training before taking up their permanent commissions with their Indian Army regiment. Officers were also looked down upon, partly because they lacked the seniority that their British equivalents had, when Indian and British units served together.

Details of officers (and soldiers) of the East India Company Army and its successors are, for the most part, held in the British Library's Asia Pacific and African Collection (APAC), formerly the Oriental and India Office Collections (OIOC) or the India Office Library and Records (IOLR). Unfortunately there are many different, and at times duplicated, record sources, so it is even more important than normal to know roughly when and where your ancestor served.

Probably the easiest source for officers is the published Bengal, Bombay and Madras Army Lists which run from 1759 (1781 Bengal) to 1889 when they were combined in the *India Army List*. These volumes are very similar to the British *Army Lists* which are described elsewhere. As well as officers from Britain they include Indian officers. The British Library has a complete set of these books, and you may find the occasional volume elsewhere including the library of The National Archives. The BL has a number of cumulative indexes to the various pre–1889 *India Army Lists*.

Service records for officers and warrant officers in the Indian Army only begin in 1900 and continue to 1950. They are to be found in series IOL L/MIL/14, and there is an index to open files in the OIOC reading room at the British Library.

These records are closed to the public until seventy-five years from the date of entry of the individual into the Indian Army. The files are opened annually on 1 January each year. Files for 1930 were thus opened on 1 January 2006: those for 1931 will be opened on 1 January 2007 and so forth.

Career details from the more recent files can be supplied, on application to the library, to the individual themselves, the next of kin or somebody who has been granted permission by the next of kin. More information can be found in a free leaflet *India Office Military Records*, which is available from the British Library.

An unusual gravestone for Captain G N Rawlings, late Indian Army, in Acton Cemetery. His military service can be traced using records at the British Library.

12.1.2 Other ranks

The British Library has muster rolls and registers of recruits for European soldiers who served with the HEIC Army. However they are arranged by Presidency, so it is helpful before you start to know with which unit he served.

12.1.3 Miscellaneous sources

The OIOC has a unique collection of ecclesiastical returns: copies of births/ baptisms, marriages and burials of European and Eurasian Christians in south Asia (including Burma, but excluding Ceylon) and other territories controlled by the East India Company, from 1698 into the 1960s. They can search for a specific return and supply you with a copy of the entry and will supply a certified copy on Foreign and Commonwealth Office headed paper for an additional fee. There is a fixed price of £35.25 (including VAT) for a search. Photocopies, including certified copies, are available for an extra charge.

12.1.4 East India Company and Indian Army material at Kew
The library of The National Archives has a selection of *Army Lists* for the Indian Army and its predecessors, together with Thomas Carnegy's *Ubique: War services of all the officers of HM's Bengal Army* (first published 1863, reprinted JB Hayward and Son, 1985) and SB and DB Jarvis, *The Cross of Sacrifice: Officers Who Died in the Service of the British, Indian and East African Regiments and Corps, 1914–1919* (Naval and Military Press, 1993), which describes Indian Army officers who lost their lives during the First World War.

In the records at Kew are lists of officers of the European regiments, 1796–1841, in WO 25/3215-3219.

Registers and indexes of East India Company Army pensions between 1849 and 1876, and Indian Army pensions from 1849 to 1868, are in WO 23/17-23. A register of pensions paid to former soldiers serving with the East India Company between 1824 and 1856 is in WO 25/3137, with an index online at http://members.ozemail.com.au/~clday/pensioners.htm.

Men from Indian regiments who were awarded the India General Service Medal and other medals appear in WO 100. They include NCOs and other ranks, who were normally native Indians.

War diaries of Indian Army units which served under direct British command during the First World War are in WO 95. War diaries for the Second World War, again for units under direct British command, are in series WO 169 and WO 172.

12.1.5 Further reading
The best guide to these records is Ian A Baxter *Baxter's Guide: Biographical Sources in the India Office Records* (3rd Edition, Families in British India Society, 2004). The British Library has a free leaflet *India Office Military Records* which can be obtained from the Library and is available in the reading rooms. A more detailed description of the sources can be found in Anthony J Farrington, *Guide to the Records of the India Office Military Department* (London, 1982).

Two useful books are VCP Hodgson, *Lists of Officers of the Bengal Army* (London, 1927–1928, revised 1968) and Byron Farwell, *Armies of the Raj* (London, 1990) which is a social history of the Indian Army. A three-volume set packed with detailed, comprehensive data about every one of the Indian Army's battalions, brigades, and divisions during the Second World War is Chris Kempton, *Loyalty and Honour: The Indian Army, September 1939 – August 1947* (Military Press, 2004).

There are a number of books about the experiences of the British in India, both in the Indian Army and in the British Army garrisons. An excellent recent account is Richard Holmes' *Sahib: The British Soldier in India* (HarperCollins, 2005). Another book, concentrating on the officer corps, is David Gilmour, *The*

Ruling Caste: Imperial Life in the Victorian Raj (John Murray, 2005). Based on interviews with the men and women who served in India is Charles Allen, *Plain Tales from the Raj* (Abacus, 1988). The most atmospheric source is undoubtedly the poems and short stories of Rudyard Kipling, although historians disagree about how accurate a portrayal of soldiering they were.

There are several useful websites which provide help and resources, particularly the ones maintained by the Families in British India Society (FIBIS) at www.fibis.org and Cathy Day http://members.ozemail.com.au/~clday/index.html An Indian account of the history of the Indian Army can be found at http://indianarmy.nic.in/arhist.htm. Histories of the Indian armed forces during the world wars are at www.mgtrust.org.

Descriptions of the major sources can also be found on the British Library's website www.bl.uk/collections/oiocfamilyhistory. Catalogues to many former India Office series of records can be found on the Access to Archives database, www.nationalarchives.gov.uk/a2a.

12.1.6 Address
British Library Asia Pacific and African Collection
Oriental and India Office Collections
96 Euston Rd, London, NW1 2DB
Tel: 020-7412 7838, email: oioc-enquiries@bl.uk, www.bl.uk.

12.2 British Army in India
Service in India was generally popular, if for no other reason than the pay went further and the natives could generally be ill-treated without reprisal.

Many men, however, suffered badly in the climate, particularly before the introduction of modern medicines. Private John Hardy was discharged from the Royal Horse Artillery in 1877 after fifteen years service, as a result of:

> *General debility resulting from an accident to the head from being pitched from a gun on parade in India. Has induced vertigo. Affected by the climate of India. Will be able to contribute to his own support in some such quiet occupation. (WO 97/1797)*

Service Records of officers and men generally, for the British Army in India as for all others, are described elsewhere. However, The National Archives has a few records particularly relating to British forces in India.

A list of British officers who served in India between 1796 and 1804 is in WO 25/3215. Records for soldiers discharged on return from India before 1806 will be found in the depot musters of their regiments. Between 1863 and 1878 the discharges of men returning from India are recorded in the musters of the Victoria Hospital, Netley (WO 12/13077-13105); between 1862 and 1889 similar information is in the muster rolls of the Discharge Depot at Gosport (WO 16/2284, 2888–2915). Except for a few eighteenth century Artillery rolls, there are no musters of artillery and engineers in India but musters of infantry and cavalry regiments in India between 1883 and 1889 are in WO 16/2751-2887.

An incomplete series of nineteenth century regimental histories of British regiments in India can be found at http://members.ozemail.com.au/~clday/regiments.htm.

12.3 Other sources

Registers of garrison churches, and other churches used by soldiers and their families, are held by the diocesan authorities in India. Births, marriages and deaths for officers and men of the British Army in India appear in the Chaplain's Returns held at the Family Records Centre.

The National Army Museum holds Hodson's Index, a very large card index of British Officers in the Indian Army, the Bengal Army and the East India Company Army, but not the British Army in India. Many of the entries go beyond bare facts to include colourful stories of life. Civilians and government staff are included if they had seen Army life.

The Society of Genealogists, 14 Charterhouse Buildings, Goswell Road, London, EC1M 7BA, Tel: 020-7251 8799, (www.sog.org.uk) has a number of collections and lists relating to the British in India including material relating to individual officers.

The Gurkha Museum, Peninsula Barracks, Romsey Road, Winchester, SO23 8TS (www.thegurkhamuseum.co.uk) has material relating to these units.

The British Empire and Commonwealth Museum, Clock Tower Yard, Temple Meads, Bristol, BS1 6QH (www.empiremuseum.co.uk) has a number of collections of private papers relating to the Indian Army, some regimental magazines and other printed material, as well as some 2 million photographs for the Empire in general. A detail research guide can be downloaded from their website.

Chapter 13

DOMINION AND COLONIAL FORCES

From the late nineteenth century, units were raised across the self-governing colonies of Canada, Australia, New Zealand and South Africa, partly to relieve the Imperial defence budget but also in patriotic response to crisis within the Empire.

Some volunteers from New South Wales served in the Maori wars in New Zealand. Nearly 800 men and officers from New South Wales also served during the Sudan War in 1885. A list of these volunteers can be found at www.awm.gov.au/database/sudan.asp. In addition 395 Canadian boatmen were recruited to help take the British expedition's boats through the Nile rapids.

The first war, however, in which forces from the colonies played a significant role in an Imperial adventure, was the Boer War between 1899 and 1902, where nearly 30,000 men from Australia, Canada and New Zealand and a further 52,500 from South Africa itself saw service. This is discussed in more detail in Chapter 14.

Grave of the Australian Captain Clarence Smith Jefferies VC at Tyne Cot cemetery near Ypres. He was killed in action during the Battle of Passchendaele and was posthumously awarded the medal for attacks on enemy machine-gun posts.

During the First World War the majority of men in the dominion armies had been born in Britain and were responding to the call of the mother country. As so many service records and other resources are available online they are easy to check for long lost ancestors who perhaps served in a Canadian or Australian unit. Some British residents also managed to join up, largely because pay and conditions were generally better and discipline less strict (something which particularly irked senior British commanders). Even so infantry units built up a reputation as canny and courageous fighters. This is reflected in the casualties suffered, which were as great if not greater than for British. Of all the countries involved in the Great War the one which proportionately suffered the worst casualties was New Zealand.

The decline of emigration between the wars and the use of conscription (both Canada and Australia had rejected conscription during the First World War) meant that dominion forces during the Second World War were now largely made up of native-born citizens.

One effect of the two world wars was to create nation states out of once compliant colonies – in part this was because the politicians wanted a greater say in the running of the war, but also because of concerns over poor British military leadership. This is reflected in the Australian and, to a lesser extent New Zealand, nationalism which still revolves around the ANZAC landings on Gallipoli on 25 April 1915 and the sacrifices made by Australian and New Zealand soldiers on the beaches.

Even today more is made of the traumatic experiences of the two wars in Canada, Australia and New Zealand which is reflected in some superb online resources, particularly the Australian War Memorial www.awm.gov.au.

Men from the Ceylon Tea Planters' Battalion play cricket at a rest camp in Egypt in 1917. (Taylor Library)

The other colonies, with the exception of India, played less important roles. In part this was because they were largely undeveloped, but undoubtedly racism played a part.

In the eighteenth and nineteenth centuries troops from the West Indies (under white officers) served in West Africa and other West Indian islands, where it was thought that they would be better able to cope with tropical conditions.

During the First World War, two battalions of the West India Regiment who served in France were largely employed in labour duties, and obstacles were placed in the way of West Indians who wanted to enlist in British regiments. During the Second World War, Nigerian troops fought bravely in Burma, and East African soldiers served during campaigns in Africa. Only six Victoria Crosses have been awarded to servicemen from the colonies, three of them awarded during the Second World War.

The records

13.1 General resources
The graves of men from the Commonwealth and Empire who died during the two world wars and other conflicts of the twentieth century are maintained by the Commonwealth War Graves Commission (www.cwgc.org). For more details see Chapters 15 and 16.

The National Archives has copies of war diaries for some units, generally those operating under British direct command overseas.

The promotion of officers and the award of gallantry medals are recorded in the *London Gazette* in the same way as for the British Army. TNA has a complete set of *Gazettes* and they can be searched online at www.gazettes-online.co.uk.

An interesting site explaining the role of Indian and colonial forces during the two world wars is at www.mgtrust.org.

13.2 Australia
Before the establishment of the Commonwealth of Australia in 1901 defence was the responsibilities of the colonies. You will need therefore to contact the appropriate state archive. Details of the State Records of New South Wales can be found at www.records.nsw.gov.au/staterecords.

Service records for conflicts from the Boer War to the Second World War and after can be obtained from the National Archives of Australia. Full details are given on their website www.naa.gov.au or you can write to National Archives of

Australian troops cheer a passing general, 1916. (Taylor Library)

Australia, Defence Service Records, PO Box 7425, Canberra BC, ACT 2610. The current charge is AU$25 for a set of records.

The Australian equivalent of the Imperial War Museum is the Australian War Memorial (AWM), which has a superb collection of material relating to Australian forces since 1901. Their website has many online databases containing details of 'diggers' who served during the two world wars, including rolls of honour and lists of men and the units they served with, together with a mass of other information about the Australian Army and its history. The website address is www.awm.gov.au or you can write for information to the Australian War Memorial, GPO Box 345, Canberra, ACT 2601.

13.3 Canada

Online databases for Canadians who served in the Boer and First World Wars can be found at www.collectionscanada.ca/genealogy/022-500-e.htm together with the scanned images of their attestation papers, which give basic personal information. Also available online are scanned images of war diaries compiled by the Canadian Expeditionary Force between 1914 and 1918.

It is possible to order copies of full service records from the Personnel Records Unit, Library and Archives Canada, 395 Wellington Street, Ottawa, ON, K1A 0N3.

Service records for Second World War Canadian soldiers are not yet open. Access is possible only to the files of individuals who died twenty or more years ago. You will need to write, with details of the individual's full name, date of birth and service number (or social security number) to the Personnel Records Unit at the address above. If the individual died after leaving the services you will also need to prove his or her date of death.

The Canadian Virtual War Museum, www.vac-acc.gc.ca/general/sub. cfm?source=collections/virtualmem contains details of Canadians who died in the world wars and other conflicts, with their last resting place. The information is very similar to that supplied by the Commonwealth War Graves Commission. In addition there are also pages about Canada during the two world wars.

During the First World War, Newfoundland provided the Royal Newfoundland Regiment, which was decimated on 1 July 1916 at Beaumont-Hamel where there is now a moving memorial. Records are held by the Provincial Archives, 9 Bonaventure Avenue, PO Box 1800, Station C, St John's, NL, A1C 5P9 www.therooms.ca/archive. The website contains an exhibition about the Regiment during the First World War.

The Canadian Military Heritage Project www.rootsweb.com/~canmil/ancestor.htm contains pages on all conflicts in which Canadians were involved, including a useful page offering advice about tracing military ancestors.

13.4 New Zealand
As in Britain, personnel records of men who served in the New Zealand Expeditionary Force are split. Archives New Zealand has records for men who served in the Boer War and the First World War. Their details are: Archives New Zealand, PO Box 12-050, Wellington www.archives.govt.nz.

Service records for later conflicts are held by the New Zealand Defence Force, Personnel Archives, Trentham Camp, Private Bag, Upper Hutt www.nzdf. mil.nz/personnel-records/archives.htm. A charge may be made for supplying service records.

Archives New Zealand also has operational records of the Army including war diaries.

Many unofficial records including unit records, newspapers and personnel papers, are kept by the Army Museum, PO Box 45, Waiouru, www.armymuseum. co.nz.

13.5 South Africa
Service records are held by the Department of Defence Documentation Centre, Private Bag X289, Pretoria 0001. Requests can be emailed to sandfdoc@ mweb.co.za. The centre houses the official records of the Department of Defence

King George V inspects New Zealand troops on Salisbury Plain. (Taylor Library)

as well as a collection of unique publications, unit history files, war diaries, photographs, maps and pamphlets pertaining to the Army. Some other material may be available from the National Archives of South Africa, Private Bag X236, Pretoria 0001, www.national.archives.gov.za.

The South African National Museum of Military History has many records relating to South African involvement in conflicts as well as displays of artefacts and weapons. The website also has useful information about tracing South African soldiers. Its address is PO Box 52 090, Saxonwold 2132, www.militarymuseum.co.za.

Further reading
Most websites cited above contain useful leaflets about tracing ancestors who were soldiers in Commonwealth and Empire armies. Two books which contain more information are:

Simon Fowler, *Tracing First World War Ancestors* (Countryside Books, 2003)
— *Tracing Second World War Ancestors* (Countryside Books, 2006)

Chapter 14

THE SOUTH AFRICAN WAR (1899–1902)

he Second South African War, universally called the Boer War, was
the most difficult war engaged in by the British Army in the 100 years
between the Battles of Waterloo and Mons. In part this was due to a wily
opponent armed with the latest European weaponry and a superb knowledge of
local conditions. But it did not help that it took the Army months to adjust to
the realities of fighting on the veld after a number of high profile reverses,
notably the sieges at Ladysmith and Mafeking.

In effect there was a war of two halves. The nine months between the
declaration of war in October 1899 and the occupation of Pretoria in July 1900
was a period of conventional warfare. Much of the remaining two years was
devoted to tracking down small bands of Boer mounted fighters, known as
Commandos. This led to the infamous concentration camp policy of housing
Boer women and children in camps where they could not provide succour to
their menfolk.

A total of 448,895 soldiers from Britain and the Empire served in South
Africa during the war, of whom 256,340 were regular soldiers or reservists
(that is men who had formerly served in the Army and were still liable to be
called up), 45,556 Militia, 36,553 Yeomanry (cavalry) and 19,856 volunteers.
This was the biggest and most varied force to see action between Waterloo and
the First World War.

The total number of casualties from all causes, including men wounded, was
52,156. There were 20,721 deaths, of which only 7,582 had been killed in action
or died of wounds and 13,139 (63 per cent) had died from disease (generally
enteric or typhoid fever). This was the last war in which more men died of
sickness than from enemy action.

Obverse and reverse of Queen's and King's South Africa medals.

14.1 Service and Pension Records

14.1.1 The regulars

Service records for men who served in the Boer War are in several sources. Records for soldiers in the Regular Army are in WO 97. A number of men also served during the First World War, so you might need to check these records as well (see Chapter 15).

There are several series of records relating to subsequent pension claims. The two most promising sources are WO 148, for claims made between 1901 and 1904, and PMG 9/48–50 for later claims. Occasionally, individual claims can be found in the selected war pensions award files in PIN 71.

It is fairly easy to trace the career of an officer from the *Army Lists*. Service records of officers are for the most part in WO 76 with some records, especially for men who saw service in the First World War, in WO 339 and WO 374 (with indexes in WO 338). Embarkation returns for officers are in WO 25/3520–3522.

Short biographies of officers who lost their lives can be found in Mildred G Dooner, *The "Last Post": being a Roll of all Officers (Naval, Military or Colonial) who gave their lives for their Queen, King and country in the South African War, 1899–1902* (Simpkins, 1903, reprinted JB Hayward and Son, 1980).

A list of officers present during the siege of Ladysmith is in WO 32/7114B.

14.1.2 Territorial forces

One unit particularly captured the imagination: the Imperial Yeomanry which aimed to provide mounted infantry troops. Initially the idea was to recruit

Photographs of volunteers from Angus who enlisted to fight in South Africa.

men from the land-owning class; men who were used to riding and hunting. The early months saw, for example, a unit of Irish huntsmen, including the Earl of Longford and Viscount Ennismore, join the 13th Battalion, and the City Imperial Volunteers (CIV) was entirely supported by donations from firms and individuals in London. The second and third contingents, recruited when it became clear that victory would be a drawn out and tedious affair, were much more proletarian in nature. Recruits were motivated less by patriotism than the promise of 5s a day.

Eventually some 40,000 officers and men enlisted. Recruits had to be between the ages of twenty and thirty-five, of good character, who could ride and shoot. Initially men

Driver William Robertson, Imperial Yeomanry. The Boer War was very mobile and horses and good horsemanship were at a premium.

were to provide their own horse, clothing and saddle and the government would provide guns, ammunition and equipment, although this requirement was dropped in 1901.

Records for the Imperial Yeomanry are with The National Archives. Short Service attestation papers are in WO 128, arranged by service number (with an index in WO 129). The records will tell you where the man was recruited, address, next of kin and in which unit and which parts of South Africa he served.

No records for officers are known to survive, although there is a nominal roll in WO 129/12. It may be worth checking the records for the First World War (see Chapter 15) as many officers re-enlisted in 1914.

There are further rolls of Yeomanry officers, non-commissioned officers and men, organised by unit, in WO 108/93-95, and you may find additional details here.

Medal rolls for the two campaign medals are listed in WO 100/120-130, 356–357. Awards of the Imperial Yeomanry Long Service and Good Conduct Medals were announced in Army Orders, now in WO 123. Casualty books listing when and where a man died are in WO 129/8-11.

Kevin Asplin has recorded the names of all 40,000 Imperial Yeomen and members of associated units at http://hometown.aol.co.uk/kevinasplin/home.html. He has also produced a book *Roll of the Imperial Yeomanry, Scottish Horse and Lovat Scouts 1899–1902* (DP&G Publishing, PO Box 186, Doncaster DN4 0HN). A list of units can be found at www.regiments.org/milhist/uk/cavyeo/ImpYeo.htm.

An article about the Imperial Yeomanry appeared in *Family History Monthly* (104) May 2004.

14.1.3 Colonial units

Small numbers of troops from Australia, Canada and New Zealand also served in South Africa. Just under 30,000 came from the colonies: 16,000 Australians, 6,400 from New Zealand and 6,000 from Canada. However the greatest number, 53,400 in total came from South Africa itself. One thousand four hundred Australians, 421 New Zealanders, 507 Canadians and 8,187 South Africans were either killed in action, died of wounds or sickness or were wounded.

John Stirling, *The Colonials in South Africa 1899–1902* (1907, reprinted by JB Hayward & Son, 1990) gives details of the service of every colonial unit that took part in the campaign, including the locally based irregulars and Australian, Canadian and New Zealand troops, based on the Commander-in-Chief's despatches.

Correspondence about gallantry awards to colonial units can be found in WO 32/7614, 8182–8188.

Australian records are largely held either by the National Archives of Australia or the Australian War Memorial. There are links to online databases to nominal and casualty rolls for all Australian troops who were in South Africa at www.awm.gov.au/research/infosheets/south_africa.asp. Another excellent website devoted to the war is http://users.westconnect.com.au/~ianmac/boermain.html. The relevant Australian National Archives page is www.naa.gov.au/the_collection/family_history/armed_services.html#boer_war.

Service records for Canadian forces are at the Library and National Archives of Canada www.collectionscanada.ca. They are available online free of charge. The site also has medal registers and land grant applications made by troops who subsequently returned to Canada. There is a search facility to allow you to find an ancestor even if you are not sure of their details. The British Isles Family History Society of Greater Ottawa has indexed the records and published the results as *Index to Canadian Service Records of the South African War (1899–1902) held at the National Archives of Canada* (Ottawa, 1999).

Links to records relating to New Zealanders who served in South Africa are at www.nzhistory.net.nz/Gallery/SAW/index.html. They include lists of men who embarked for South Africa and a roll of honour for those who did not return. Service records for the Boer War are with Archives New Zealand in Wellington. Details can be found at www.archives.govt.nz/doingresearch/nzdfpersonnelfiles.html.

There were also a number of irregular South African formations. Ostensibly made up from (white) South Africans, they also included many British – in part because it was easy to switch regiments for the bounty money. They had names such as The Commander-in-Chief's Bodyguard, The Umvoti Rifles, The Cape Town Highlanders and The Transkei Mounted Rifles. A full list, together with brief histories, are given in John Stirling, *The Colonials in South Africa, 1899–1902* (JB Hayward and Son, 1990).

Enrolment forms for local armed forces are in WO 126 with nominal muster rolls in WO 127. Medal Rolls are in WO 100, arranged by unit. Kevin Asplin's website (see above) has muster rolls for a few units.

14.1.4 Women in South Africa

Some women served in South Africa as nurses. No service records survive, but they are listed in the Royal Army Medical Corps medal rolls (WO 100/229, 353). However the rolls only list female nurses recruited locally (about a third of the total number) who worked in hospitals which were approved for the reception of British wounded. Some material about the award of pensions is in WO 23/181.

A few women were awarded the Royal Red Cross Medal for bravery or exceptional duty; rolls are in WO 145.

More information can be found in Anne Summers *Angels and Citizens. British Women as Military Nurses 1854–1914* (Routledge and Kegan Paul, 1988) and especially Sheila Gray, *The South African War, 1899–1902: Service Records of British and Colonial Women. A Record of the Service in South Africa of Military and Civilian Nurses, Laywomen and Civilians* (Sheila Gray, 1993), which contains details of 1,700 of the 2,000 or so nurses who were in the theatre of operations.

14.2 Medals

14.2.1 Campaign

There were two campaign medals for the War: the Queen's South Africa Medal (QSA) awarded to men who served in the campaign up until the Queen's death on 22 January 1901, and the King's South Africa Medal (KSA) for men who served after this period. Some 180,000 QSA medals were issued, with rather fewer KSA medals. Rolls for the QSA are in WO 100/112-301, 371, 381–383 and the KSA in WO 100/302-367, 369–370.

The Queen's Mediterranean Medal was awarded to volunteer and Militia troops who had replaced their regular counterparts in garrisons across the Mediterranean allowing more regular troops to be available for the South

African War. The medal was also awarded to troops who guarded Boer prisoners of war at the POW camp on the island of St Helena. Approximately 5,000 medals were awarded in total and the medal roll is in WO 100/368.

An article about the medals appeared in *Ancestors Magazine* April 2005 (32).

14.2.2 Gallantry

Gallantry awards can be found in several places. Registers of awards of the Victoria Cross are in WO 98/4, 7 with correspondence in WO 32 (Code 50). In all seventy-eight VCs were awarded during the war. They are described online at www.victoriacross.net.

Other ranks were entitled to the Distinguished Conduct Medal (DCM). Recipients are listed in PE Abbott, *Recipients of the Distinguished Conduct Medal, 1855–1909* (JB Hayward and Son, 1975).

South African War Honours and Awards 1899–1902 (Arms & Armour Press 1971), is a reprint of an original 1902 book published by the *Army & Navy Gazette* and lists all the gallantry and good conduct awards that had appeared in the *London Gazette* during the war.

Many men were mentioned in despatches for their heroism or meritorious conduct during the campaign. Their names were recorded in the despatches sent to London and printed in the *London Gazette*. There is no other list and recipients were not awarded a medal or other symbol to wear on their uniform notifying the award. Their details are published in *South African War Honours and Awards, 1899–1920: Officers and Men of the Army and Navy Mentioned in Despatches* (Greenhill, 1987).

14.3 Casualties

14.3.1 Deaths

A complete list of men who died or who were wounded in action is in WO 108/360 and it has been published as Frank and Andrea Cook, *South Africa Field Force Casualty List, 1899–1902* (London, 1972). This is arranged by unit rather than by name, so it may be easier to use Alexander M Palmer's *The Boer War Casualty Roll, 1899–1902: An Alphabetical Listing* (Military Minded, 1999), which lists casualties alphabetically giving such details as the unit they were with, what type of casualty they were (killed, wounded, captured, died of wounds), place and date they became a casualty.

For men who served in Natal, check *The South African War Casualty Roll: The 'Natal Field Force' 20th Oct. 1899–26th October 1900*.

Other lists of casualties are in WO 108/89-91, 338 and WO 25. A list of soldiers buried in colonies in Cape Province is DO 119/1479.

Soldiers' Effects Ledgers from April 1901 are with the National Army Museum – earlier ones have been lost. They are arranged by name of soldier and the registers themselves record the full name, regimental number and rank, date and place of death and birth, date of enlistment, trade on enlistment and next of kin or legatee if a will had been made, together with the amount they were due.

The South African Heritage Resources Agency has a computerised database for all those who died in military conflicts on South African soil, including the Anglo–Boer War, which will tell you where a man is buried. Contact details: SAHRA, PO Box 87552, Houghton, 2041, South Africa, email: jbeater@jhb.sahra.org.za.

14.4 Operational Records

It can be almost impossible to trace what individual units did as the invaluable habit of writing a daily unit war diary did not start until after the Boer War was over. Probably the best place to start are the published regimental histories.

The Grenadier Guards at Biddulphsberg.

Staff diaries may help. Unfortunately their survival rate is poor. Diaries for December 1899 and January 1900 are in WO 32/7868, 7876–7877. Those for the Natal Campaign for May and December 1900 are in WO 32/7901, 7951.

The South African War Papers, in WO 108, consists chiefly of the correspondence of the Commander-in-Chief, South Africa (Lord Roberts) with reports and returns received by him from unit commanders. There are also some nice photographs of blockhouses. Similar material can be found in Lord Robert's personal papers in WO 105.

Another important source is the material in the War Office general correspondence (WO 32/7844–9119). Some intelligence reports, including reconnaissance reports, are in WO 33/152–198.

Maps, drawn up for use by the staff in planning campaigns and operations, are in series WO 78.

14.5 Other Sources

14.5.1 Photographs and ephemera

Under copyright legislation of 1870 anybody who wished to copyright an image had to submit a specimen to Stationer's Hall. This material is now in series COPY 1. The great interest in the War felt by people in Britain is vividly expressed in these collections. There are thousands of photographs, posters, point of sale material, brochures, chocolate wrappings etc. containing images from the war to be found here. The one problem is that there is no index at all, so it can be difficult to easily find a specific image.

Further reading

More about tracing ancestors who fought in the Boer War can be found in Phil Tomaselli, *The Boer War 1899–1902* (Federation of Family History Societies, 2006).

There are a number of books about the Boer War. The best remains Thomas Pakenham, *The Boer War* (Weidenfeld and Nicolson, 1979 and widely available in paperback). Also useful are Denis Judd and Keith Surridge, *The Boer War* (John Murray, 2002) and Michael Carver, *The National Army Museum Book of the Boer War* (Pan, 2000).

Chapter 15

THE FIRST WORLD WAR

T he First World War was the first war which engaged almost every able-bodied man and woman and affected everybody else young and old, not just in Britain but in the British Dominions, France, Germany, Russia and the less important combatants.

After a few weeks in August and September 1914 fighting soon became a war of attrition based a network of trenches which snaked nearly 500 miles from Nieuport on the Belgian coast to the Swiss border.

It took three years, and the loss of hundreds of thousands of men, before the British commanders could begin to work out a successful strategy. This has led to the myth of 'lions led by donkeys', popularised in lazily researched books and TV series ever since.

In their defence the British commanders were physically isolated from the fighting – even if they had been present the battle was so vast that they would have been none the wiser – and relied on imperfect communications and intelligence. That said, with hindsight, the tactics were often imperfect and sometimes pure wrong-headed.

It was only in the latter half of 1917 that the strategy which would win the war (admittedly against an increasingly weakened enemy) was perfected. The hundred days before the Armistice in November 1918 saw the greatest series of British military victories ever.

It was a war in which the Germans had two great advantages. The first is that they were defending the territory they had captured, so had little incentive to attack the enemy. Their major offensive in March 1918 was a pre-emptive attack on the British and French before American troops could arrive in France. It faltered largely because the advancing soldiers could not be re-supplied.

The second advantage is that very largely they occupied what little high ground existed particularly in Flanders, which made it even more difficult for the

The Menin Gate at Ypres; the memorial to nearly 55,000 men who fell and have no known grave in Flanders between the outbreak of war in 1914 and August 1917.

Allies to attack. This will quickly become clear if you visit the battlefields – the British campaigns around Ypres in 1915, on the Somme in 1916 and at Passchendaele in 1917 were all largely uphill.

For most soldiers the centre of their world was the battalion, which comprised about 1,000 men and thirty-five officers at full strength, although the numbers could be much reduced. It was an administrative and fighting unit, which not only prepared its members for battle, but fed, paid and clothed them, arranged their leave and tried to supervise their recreational needs. Battalions were sub-divided into four companies, sixteen platoons and sixty-four sections. Officers commanded companies and platoons, while non-commissioned officers were in charge of sections. Each battalion also had a number of specialists, such as signallers, cooks and machine (Lewis) gunners.

In the pre-war Army each regiment had comprised two battalions of regular soldiers plus several battalions of territorial or part-time soldiers. This number dramatically increased during the war; a number of regiments now had a dozen or more battalions.

Brigades were above battalions in the chain of command. An infantry brigade consisted of four battalions, machine-gun and trench-mortar units under a brigadier general.

Of greater importance was the division, which was the basic tactical unit. A division comprised about 20,000 men from infantry, artillery and engineer units as well as medical and signal services. Commanded by a major general it held a front line of approximately two to three miles. Above divisions were corps and armies and at the pinnacle was the General Headquarters (GHQ) under first Sir John French and then, from December 1915, Sir Douglas Haig.

The British occupied about eighty-five miles of trenches from just north of Ypres to the Somme River in northern France. The area around Ypres was low lying and prone to flooding (hence the dreadful mud which enveloped thousands of unwary men), while the area around the Somme was chalk down land, which was dryer and easier in which to dig trenches.

It is important to remember that there was not just one trench running north-west to south-east across western Europe, but a maze of trenches connecting and reinforcing the front line which separated the enemies, across no man's land.

In the British sector three parallel lines were the norm – the front, support and reserve trenches. They were built either below or above ground as breastworks (a low earth wall) or were part trench and breastwork combined. Usually the front line consisted of two trenches; the fire and the command trench. The fire trench was continuous, but not straight, and traversed at intervals by earth buttresses, which protected against snipers and shell blasts. Here men would be positioned manning machine guns or standing on guard, particularly at dusk and dawn which were thought to be the most likely time for enemy attacks.

About twenty yards behind was the command trench, which housed dug-outs, which gave some protection in the event of an enemy attack, and the latrines, which attracted artillery attacks, because from the air they looked like machine-gun posts.

Further behind were the support trenches, with deep dug-outs to house support troops for the front-line garrison. Further back still, perhaps half a mile behind the front line were the reserve lines, where more troops were stationed.

The front line, support and reserve lines were connected at frequent intervals by winding communications trenches, along which passed reliefs, rations, supplies of ammunition and telephone cables to battalion and battery headquarters.

In the rear were the artillery batteries, which were located far enough away to be out of range of enemy small-arms fire.

For miles behind the front line were the storage and ammunition dumps, engineering workshops and kitchens and casualty clearing stations dedicated to supporting the men in the front line. Goods were largely moved by horse drawn carts, although there was a network of light railways bringing up shells for the artillery.

Among them dwelt those civilians who had not fled the fighting. A number of accounts describe how local farmers ploughed and harvested their fields almost up to the front line. The small towns and villages had a plethora of estaminets (bar cafes) where men would forget the war for a few hours, drinking the local wine (referred to as 'plonk'), eating egg and chips and ogling the owner's wife or daughters. There was also a network of official and semi-official canteens – run by the Army, the YMCA or other charitable bodies – where men could write home, play cards or read.

Over 5 million men served in the British Army; almost exactly half had volunteered and the other half were conscripts. Inevitably this transformed the Army from an insular, professional force to one largely comprising civilians in uniform.

A million men enlisted in what became known as the New Army in the early months of the war. Most joined up in a fit of patriotic fervour. Recruits had to be between the ages of eighteen and forty-five and taller than 5 feet 6 inches (although a number of bantam battalions were raised for men shorter than this). There is plenty of evidence that the Army connived in the recruitment of under-age (and some over-age) soldiers, although it is most definitely not true that the typical Tommy was sixteen and lied about his age to get in!

By the end of 1915, nearly 2.5 million men had joined up, many into 'Pals' Battalions, organised on a local basis. Battalions such as the Hull Commercials shared an occupation; others, like the Glasgow Tramways Battalion, shared an employer; the Tyneside Irish had a common background. The disadvantage was that if a unit suffered heavy casualties, as happened with the Accrington Pals at the Somme, it could have a devastating effect upon a community.

Numbers volunteering quickly fell away as the horrors of war became known or because men felt unable to leave their families and work. After considerable debate, conscription was introduced in March 1916, initially only for bachelors between the ages of eighteen and forty-one, but it was quickly extended to married men as well. Men who were doing war work in factories and mines were placed in reserved occupations and exempted from service, although as the demand for manpower increased these occupations were 'combed out' (as the phrase went) to release men to the Army.

There were two aspects to the training provided. The first was to provide the skills needed to survive in the trenches, such as dealing with gas attacks or using bayonets. The other was to inculcate discipline and to break any civilian individuality, in order to ensure that orders were implicitly followed and also to install pride in the regiment. For many men it was perhaps the first time in their lives that they had had a decent diet and been fit.

The period of basic training varied – initially it could take up to eighteen months, but with the demands to fill the trenches this was progressively reduced. By 1918 this might be as little as six weeks. After this period, specialists such as drivers or gun layers, would depart for further training in their specialities. The infantry would be sent for supplementary training in the skills of trench warfare.

After a short period of embarkation leave, the men would depart for France, where there was further training, often at brutal camps such as Etaples (near Dieppe) where there was a mutiny against the conditions in 1917. Finally, followed the inevitably slow train journey to a base camp close to the front line.

Initially officers came from the same public school, middle class background that they had always come from. Of the 600 old boys from Bury Grammar School in Lancashire who served in the Army between 1914 and 1919, ninety-seven were killed in action or died of wounds.

In the early days of the war officers were often local businessmen or landowners, or perhaps had served in the pre-war Army or Militia. Training was pretty minimal, it was assumed that one's social position and the ability to give orders was sufficient. One new officer attended a course in Belfast: 'I do not think we learnt much that was useful, but the course provided an opportunity for one or two pleasant tea parties.'

It helped that pre-war many public schools and universities maintained an Officer Training Corps (OTC) which trained boys in the military service. By selecting public schoolboys, the Army ensured that its officer recruits had a consistent standard of basic training and a belief system relevant to military service. During the war many cadets graduated direct from the Corps to the front line and an increasing number of men were promoted from the ranks. Initially they came largely from the middle classes who had enlisted as privates in the enthusiasm of the early days of the war. However, as the war progressed, an increasing number of men from the lower middle classes and even the working classes were commissioned to address the shortage in officers. Often they had been corporals and sergeants, who had been recommended by their commanding officers. Because of the hidebound nature of the British class system many found it difficult to adjust, and *Punch* is full of mean-spirited cartoons about these 'temporary gentlemen' as they were sometimes called.

Officers and other ranks led surprisingly separate existences. Fraternisation between the ranks was discouraged, although inevitably there was some particularly with the democratisation of the Army in 1917 and 1918. There were separate cafes, bars and even brothels of superior quality for officers. Perhaps the only place on the Western Front where there was complete equality was Talbot House in Poperinghe near Ypres, where the rule 'abandon rank all ye who enter here' was strictly enforced.

Even so relations between junior officers and the companies and platoons could be warm. Men respected officers who knew their mind and showed leadership. Second Lieutenant Norman Collins remembered preparing to go over the top: 'Knowing one's duty took one's mind off the horrible things. You are very aware of the example you are setting the men and if they saw you flunking it – showing fear – they wouldn't think much of you.'

In turn officers often became fond of the soldiers in their charge. It was very likely that, for many, this was the first time that they had dealt with working men to any degree. The poet Wilfred Owen of the Manchester Regiment described the Wigan colliers who were members of his platoon: 'The generality are hard headed, hard headed miners, doggish, loutish and ugly (but I would trust them to advance under fire and to hold their trench).'

15.1 Service Records

15.1.1 Officers

Surviving service records for officers are at The National Archives. However, if a man remained an officer after April 1922, then his service record will be with the Ministry of Defence's Army Personnel Centre, Historic Disclosures, Mailpoint 400, Kentigern House, 65 Brown St, Glasgow, G2 8EX, Tel: 0141-224 2023, email apc_historical_disclosures@btconnect.com.

Records of Guards officers are still with the Guards Museum, Wellington Barracks, Birdcage Walk, London, SW1E 6HQ.

Officers are listed in the *Army Lists*, which were published quarterly during the war. For intelligence reasons the information given is not as comprehensive

Attestation form for Wilfred Owen, enlisting as a private into the Artist's Rifles. He later became an officer in the Manchester Regiment and was killed during the last week of the war. (TNA:PRO WO 138/74) as in peacetime lists. Even so it should be possible to track down promotions and the date they were made.

Some 271,800 officers' service records are with the National Archives, perhaps 85 per cent of the total. They are a duplicate series of folders compiled by the government, generally containing correspondence about a man's death, or pension eligibility, rather than about an individual's war service. In fact there are two series of records: WO 339 and WO 374. The latter consists of a file of Territorial Army officers, officers who came out of retirement, and other officers recruited because of their skills in civilian life (such as railway managers). These series are arranged alphabetical, so it is easy to find the file for an individual using TNA's online catalogue, particularly if you are looking for an unusual name. Otherwise you will have to use the index in WO 338.

Service files for a few notable individuals, including Field Marshal Lord Haig and the poet Wilfred Owen, are in WO 138. Biographical details of officers in the Army Medical Corps are listed in Sir William Macarlane's *A List of Commissioned Medical Officers of the Army, 1660–1960* (Wellcome Library, 1968).

15.1.2 Other ranks: service records
Unfortunately perhaps 65 per cent of service records have been lost. There appears to be no rhyme or reason to what survives or has been destroyed.

If your ancestor stayed in the Army after the end of 1920 then his service record will be with the Ministry of Defence's Army Personnel Centre, Historic Disclosures, Mailpoint 400, Kentigern House, 65 Brown St, Glasgow, G2 8EX, Tel: 0141-224 2023, email apc_historical_disclosures@btconnect.com.

Records of Guardsmen are with the Guards Museum, Wellington Barracks, Birdcage Walk, London, SW1E 6HQ.

Service records should tell you when a man enlisted and was discharged, who his next of kin was, promotions through the ranks, details of units served in, as well as medical treatment received and appearances before courts martial. There may also be correspondence about pensions, medal claims or relatives seeking a missing man. What the records will not do is tell you very much about any fighting he was engaged in; for this you will need the war diaries. These records are at The National Archives on microfilm, although it is likely they will be online within the foreseeable future. It is also possible to order films to be read at LDS Family History Centres for a small fee. They are arranged in strict alphabetical order by surname and then by forename. You should know in which unit a man served and his regimental number, particularly for men with common names. There are also a small numbers of miscellaneous files found during filming which have been placed at the end of each series.

There are three series of service records:

WO 363 sometimes known as the 'burnt' records, as they were damaged by fire or water in 1940. They relate to soldiers who were:
- Killed in action.
- Died of wounds or disease without being discharged to pension.
- Demobilised at the end of the war.
  The 'burnt documents' that survived the bombing in 1940 consist of about 20 to 25 per cent of the original total.

WO 364 originally called the 'unburnt' records, because they were either untouched by the fire or subsequently added from other sources. These service records are for:
- Regular soldiers discharged at the end of their period of service. Men who signed up for the duration of the war did not get pensions; instead they got a gratuity on demobilisation, and will not be found here unless they received a pension on medical grounds.
- Men discharged on medical and associated grounds, including those who died after the award of a pension.

WO 400 Service records for men of the Household Cavalry, which included the Life Guards, Royal Horse Guards, and Household Battalions. These records are complete.

15.2 Medals and Awards

15.2.1 Campaign medals

Every serviceman and woman who saw service overseas was entitled to two campaign medals – the British War Medal and the Victory Medal. In addition, men who were in France and Flanders between 5 August and 22 November 1914 were awarded the 1914 Star (often called 'The Mons Star') and men who served overseas between 5 August 1915 and 31 December 1915 were entitled to the 1914/15 Star. Two other campaign medals were also awarded to soldiers – the Territorial Forces War Medal (for members of the Territorial Forces at the beginning of the war who saw service overseas) and the Silver War Badge (for men who were discharged because of wounds or sickness).

The awards of campaign medals are recorded on Medal Index Cards. They are available either online at www.nationalarchives.gov.uk/documentsonline or on microfiche in the Microfilm Reading Room at Kew. They list everyone who was entitled to a medal, which is doubly helpful if the Army service record you are looking for has been destroyed. In particular they can tell you rank, regimental number, unit served in, other medals awarded, date of discharge, the theatre of war, and the dates of overseas postings. There are often other references particularly to the despatch of medals to widows and next of kin.

The medal information card for Henry P S Crozier, the author's great-uncle.

The website contains a free index, which will provide details of regimental number, regiment and rank at the time of the individual's discharge or when the medals were issued. To see the card itself, you will have to pay a fee (at time of writing £3.50).

15.2.2 Gallantry medals

As the name suggests gallantry medals, such as the Victoria Cross or the Military Medal, were awarded for individual heroism. In most cases it is not possible to find out very much about why such a medal was awarded, as citations were rarely published. In many cases medals were given out almost randomly to members of a platoon or company, which had seen action. As men in the front line cynically said, they had come up with the rations.

Details of all awards were published in the *London Gazette*, sometimes with a citation or short description of why the medal was awarded. At the very least you will get the man's name, service number, rank, regiment and the date when the award was made.

A fully indexed edition of the *London Gazette* is available online at www. gazettes-online.co.uk and is fully indexed. The National Archives has copies (in series ZJ 1) together with indexes in the Microfilm Reading Room.

The most famous gallantry medal is, of course, the Victoria Cross. Biographies of many of the 633 winners of the award during the war, together with descriptions of their exploits, are described in an excellent series of books by Gerald Gliddon (and other authors) published by Sutton Publishing. There are also several websites devoted to VC winners, of which www.victoriacross.net is the most comprehensive. Less satisfactory short biographies also appear in Wikipedia (http://en.wikipedia.org/wiki/Main_Page). A register of VC recipients can be found in series WO 98, together with copies of citations and other information.

The Distinguished Service Order (DSO) was normally only awarded to senior officers, while the Military Cross (MC) was awarded for acts of bravery to officers of the rank of captain or below. A register, arranged by date, appeared in the *London Gazette* and can be found in WO 390. Annotated copies of the *Gazettes* with the place and date where the medal was won are in WO 389. Armed with this information you can then find out more by using war diaries. Both series are only available on microfilm in the Microfilm Reading Room at Kew.

The Distinguished Conduct Medal (DCM) was awarded to non-commissioned officers and other ranks. *Gazette* books are in WO 391 and recipients can be found in the Medal Index Cards.

Recipients are also listed in RW Walker, *Recipients of the Distinguished Conduct Medal, 1914–1920* (Midland Records, 1981).

The most common award was the Mention in Despatches (MiD). During the war just over 2 per cent of the men in the British forces (141,082 officers and other ranks) were so honoured. At the time the only reward was a mention in the *London Gazette*, but subsequently King George V authorised the issue of a commemorative letter to holders and men were entitled to wear a stylised bronze oak leaf on their uniforms. Men awarded the MiD appear in the Medal Index Cards. These Cards also hold details of men who were awarded the Military Medal (MM); Meritorious Service Medal (MSM); Territorial Force Efficiency Medal (TFEM); Territorial Efficiency Medal (TEM). They give the date upon which the award was announced in the *London Gazette* (sometimes by means of a numerical code) which is explained at www.nationalarchives.gov.uk/documentsonline/medals.asp#arranged or ask staff in the Microfilm Reading Room at Kew.

More about medals (both gallantry and campaign) can be found in William Spencer's *Army Service Records of the First World War* (Public Record Office, 1998) and his *Medals: a Researcher's Guide* (TNA, 2006).

15.3 Casualty Records

Of the 5 million men who served in the British Army during the war (about 22 per cent of the total male population), just over 700,000 men and officers were killed, 1.7 million were wounded, and another 170,000 taken prisoners.

In fact it could have been a lot worse. The medical services had been overhauled in the years before the outbreak of the war and proved resilient under circumstances which no one could have predicted. Provided a man could be rescued from the battlefield or front line and taken to a casualty clearing station then he had a reasonable chance of surviving. Even so one in three casualties died. A large number of Commonwealth War Grave cemeteries are located where dressing and clearing stations were based as mute testimony to the limitations of medical science at the time.

Most wounds were caused by artillery: shrapnel was a particular killer, although 30 per cent were as the result of enemy bullets.

Each battalion had a number of stretcher bearers, generally members of the battalion band, who had been taught basic first aid. It was their job to pick up and take wounded men to the Regimental Aid Post in a dugout in a support trench where the battalion medical officer (on detachment from the Royal Army Medical Corps) would perform the briefest of first aid.

If necessary (and it generally was), men would then be stretchered back though the command trenches to the Advanced Dressing Station in one of the support trenches. At busy times it was not uncommon for men to wait patiently outside the dressing stations while their severely wounded comrades received priority treatment.

Here the wounded would have their wounds tended to, be given a tetanus injection, have a Field Medical Card made out (red for seriously wounded, green for less seriously wounded) and be placed in an ambulance which would take them to a Casualty Clearing Station which was located beyond the range of enemy artillery attack. Here men would be sifted into those who were dangerously wounded and those whose injuries were less serious.

From the Casualty Clearing Stations wounded men would be sent to hospitals in the rear or back in Britain. Over 1.1 million wounded men crossed the Channel, two-thirds of whom recovered enough to return to France to see further action.

15.3.1 The records

The Commonwealth War Graves Commission was set up to commemorate the dead of the First World War and, of course, subsequent wars. The Debt of Honour Register (www.cwgc.org) will tell you where a man is buried, when he died, his rank and the unit he served with, and sometimes next of kin. The Commission also provides the same service for people who write in or telephone their offices. Their address is: 2 Marlow Road, Maidenhead, SL6 7DX, Tel: 01628-34221.

Details of deaths are also recorded in *Soldiers Died in the Great War*, which is available on CD and online at www.military-genealogy.com. The information contained about individuals varies and is by no means accurate, but will indicate where a man died, his rank and the regiment he was with. It may also tell you where and when a man enlisted and his age at enlistment. Most local reference libraries and some family history societies should have the disc.

15.3.2 War memorials and rolls of honour
After the war some 45,000 war memorials were erected in honour of men who did not return usually for a locality and regiment or other unit and sometimes in schools, churches or work places. An ancestor may appear on two or three such memorials, or in rare cases may have been missed off altogether. My great-uncle Stanley, for example, appears on at least three. Normally all you will find is the name; sometimes the rank and unit and details of gallantry medals will also be included.

The UK National Inventory of War Memorials based at the Imperial War Museum has a database of memorials for all wars. It is online at www.ukniwm. org.uk. Although it rarely lists the names, it will tell you about the location, design and possibly how and why it was built.

An incomplete list of names on war memorials can be found on another online database hosted by Channel 4 as part of a TV series which was shown in late 2005. www.channel4.com/lostgeneration is easy to search and also contains pages about war memorials and advice about tracing the officers and men who did not return.

Scotland's war dead are honoured at the Scots National War Memorial situated in the precincts of Edinburgh Castle. More information, together with details of the men who can be found on it, is on its website www.snwm.org.

A number of studies about the men who appear on local war memorials have been researched in recent years. It is worth asking in your local library to see whether one has been done for your area.

In the years after the war many companies and schools prepared rolls of honour listing employees who served or perhaps just those who had fallen. On occasion short personal biographies were included. A probably incomplete list is given in Norman Holding's *The Location of British Army Records 1914–1918*. The National Archives, for example, has a number of rolls compiled by railway companies. The Naval and Military Press has reprinted a number of rolls of honour.

There are also a number of rolls of honour online. For example, Suffolk Record Office has posted a list of men from the county at www.suffolkcc.gov. uk/sro/roh/index.html.

15.3.3 Miscellaneous sources

French and Belgian death certificates for British soldiers who died in hospitals or elsewhere outside the immediate war zone between 1914 and 1920 are in RG 35/45-69. They are arranged alphabetically by surname.

Newspapers, local as well as national, published casualty lists. In addition most regional newspapers included short biographies, sometimes with a photograph, of local men who lost their lives. Increasing shortages of paper meant that this practice was curtailed from mid-1917 onwards. An interesting summary of other printed sources can be found in a British Newspaper Library leaflet *Family History Research and British Military History, 1801–1945* available online at www.bl.uk/collections/warfare3.html.

15.3.4 Hospital records

An efficient system of dealing with casualties was quickly introduced on the outbreak of war to ferry the sick and wounded to the appropriate casualty clearing station or hospital in the rear. A record card was compiled for each man, but they were destroyed in the early 1980s, with the exception of a 2 per cent sample. The surviving cards are at Kew in series MH 106.

The records of the hospitals themselves too have largely been destroyed. Again a small number of examples can be found in MH 106. Occasionally you may find material at local record offices.

15.3.4 Prisoners of war

Nearly 200,000 British and Commonwealth prisoners of war fell into the hands of the Germans and their allies, about half of whom were captured during the last six months of the war. Conditions could be grim, in part because of deliberate mistreatment, but more often because of increasing problems within Germany itself. Many prisoners eventually depended largely on Red Cross parcels, which were collected and packed by voluntary organisations under the leadership of the British Red Cross.

It is difficult to find out very much about individual POWs, largely because detailed Red Cross registers and other records were lost in the late 1920s. Possibly the best place to start is the Foreign Office Prisoner of War Department files in FO 383, which has recently been properly catalogued for the first time, revealing major new resources. The index is available as part of TNA's online catalogue.

A list of prisoners in German and Turkish hands in 1916 is in AIR 1/892/204/5/696-698, which indicates where the prisoner was captured and when, where they were held and their next of kin. There is a published *List of Officers taken Prisoner in the Various Theatres of War between August 1914 and November 1918* (1919, reprinted by London Stamp Exchange, 1988).

Returning British prisoners were interrogated by the authorities about their experiences and a selection of these reports (with an index) is in series WO 161. Reports by officers may be found in their service records (see above).

More about the experience of POWs can be found in Richard van Emden, *Prisoners of the Kaiser: the last POWs of the Great War* (Pen and Sword, 2000) and Robert Jackson, *The Prisoners 1914–1918* (Routledge, 1989).

15.4 Discipline and Desertion

Perhaps the most controversial issue remaining from the war is the case of the 304 men who were 'Shot at Dawn', of whom 265 were convicted of desertion. To modern eyes, at least, many of these men were clearly suffering from shell shock or a mental breakdown brought about by conditions at the front.

In fact the British Army was, in general, a very disciplined and well behaved force. Alone of the major protagonists, allied or enemy, it did not suffer a major mutiny, largely because morale generally remained high throughout the war and, by comparison with France and Germany, at least casualties were low. The mutiny at Etaples in 1917 was easily ended when many of the men's grievances were met.

Most offences were the traditional ones of drunkenness, the theft of Army property and discipline. Generally punishment was fairly light. Discipline in Regular Army battalions (that is those already in existence in August 1914) was generally tighter than in the new Army units which largely consisted of civilian volunteers and conscripts. Indeed there were fewer courts martial in 1917–1918 than in the first year of the war, by which time the Army was overwhelmingly a civilian force.

A fairly serous problem, however, was desertion. In 1917 it was estimated that at any one time 20,000 men (the equivalent to a division) were on the run from their units but tight security meant that it was almost impossible to cross the Channel. Many deserters lived in abandoned buildings and ruins near the front, surviving on food they stole or were given by sympathisers.

Even so nearly 300,000 soldiers and 6,000 officers faced courts martial during the war, generally for being absent without leave or for drunkenness. Service records should indicate whether your ancestor was put on a charge. Registers in WO 90 (for men serving overseas) and WO 92 (for men on home duty), will give brief details of the offence. Details of more serious offences can be found in WO 71 and WO 93. More information is given in TNA Research Guide *British Army: Courts Martial: First World War, 1914–1918*.

The story of the men who were punished by execution is told in Julian Putkowski and Julian Sykes, *Shot at Dawn* (Pen and Sword, 1998): the book also provides references to the records. A website devoted to the campaign to secure a pardon for these men is at www.shotatdawn.org.uk.

Not everything which appeared before courts martial was deadly serious. Although the moustache was not compulsory for officers in the British Army of the Edwardian era, shaving the upper lip was forbidden. An officer was successfully court-martialled for shaving in 1916. Fortunately when the sentence passed through the hands of Lieutenant General Sir Neville Macready, Adjutant General of the British Expeditionary Force, on its way to the Commander-in-Chief for confirmation, Macready recognised the absurdity of the regulation and had it revoked.

15.4 Operational Records

Men were rotated in and out of the trenches every few days: normally they spent seventy-two hours in the front line and a period in one of the reserve trenches before returning to a billet, perhaps a disused barn or tents a few miles behind the front line for a period of rest.

In fact some battalions saw little serious action. One regimental historian reckoned that out of the 1,553 days of the war his battalion had been in 'general and active conflict with the enemy' for just twenty of them. The rest of the time his unit had been engaged in normal trench duties and at rest behind the front line. Yet they emerged far from unscathed. The same historian thought that the battalion had been totally replenished eight times over by drafts from home. Over a fifth of men and officers had been killed and half had been wounded at least once.

Despite the best efforts of GHQ and enthusiastic junior officers, life on the front line was often quiet, with only sporadic enemy bombardments and raiding parties. Going over the top was rare. One officer, Charles Carrington, estimated that in 1916 he saw action four times and led wiring or intelligence gathering parties into no man's land on another six occasions. For most men their time in the trench, as Richard van Emden says, was 80 per cent bored stiff, 19 per cent frozen stiff and 1 per cent scared stiff.

Apart from the boredom the greatest problems facing the troops were getting enough sleep and the infestation of lice and rats. The only time that men in the front line could get any sleep was in the afternoon. During the day, when the chance of enemy attack was remote, the trenches often appeared deserted with bored sentries every few yards. Matters were different at night, when full laden replacements crept up the communication trenches to relieve troops in the front line.

Night time was also when patrols might be sent out into no man's land to find out what the Germans were doing (and possibly try to kidnap a German soldier to see what intelligence he might supply) or to replace the barbed wire, which separated the two sides. Patrols were also meant to emphasis the British domination of no man's land. They could be very dangerous, particularly for the young officers who led them. It is no surprise that the rank which proportionally suffered more casualties than any other were second lieutenants.

In many sectors, such as round Ploegsteert (Plug Street to the Tommies) on the French-Belgian border, both sides operated a 'live and let live' policy, although this depended on the aggressiveness of the units stationed in the front line.

But what is remembered today are the big set piece battles, which often seem to merge into one another. After weeks of intense bombardment, when it was hoped German defences would be destroyed, tens of thousands of men would scramble over the top into no man's land and the sights of German machine guns.

Men would know a week beforehand that a push was imminent, either because they had been briefed to be prepared or through rumours. Even fortified by a double ration of rum, the moments before going over the top were apprehensive. Most did not think they would be killed, just wounded. Some preferred to climb up the trench ladders immediately while others hung back for a few seconds.

By 1917 troops were increasingly trained for attacks with scale models of German trenches and their objectives. Tanks were initially a huge success, although they had a tendency to break down or get stuck in trenches. The only time church bells were rung during the war was to celebrate the Battle of Cambrai in November 1917 when tanks were first successfully used in battle. Even so poor communications and the natural confusion of battle meant that advantages were too often not pressed home.

15.4.1 The records

Army units of battalion size or larger, on active service, were required to keep a daily record of all significant events, together with other information such as details of operations, casualties and maps. These war diaries exist for British, Dominion, Indian and Colonial forces serving overseas, but rarely for units stationed in the British Isles.

They are mainly found in series WO 95 and are an invaluable source for tracing what an individual did day by day. However the amount of detail in the diaries varies greatly and may depend on the enthusiasm of the responsible officer and the circumstances in which they were completed.

It is also unusual for ordinary soldiers to be mentioned by name – even if they were engaged in some feat of heroism. However, the lack of personal names is not necessarily important: knowing that your ancestor was there is sufficient to build up a picture of what he was engaged in.

INTELLIGENCE SUMMARY

WAR DIARY
or
INTELLIGENCE SUMMARY
(Erase heading not required.)

Army Form C. 2118.

11TH BATTALION LEICESTERSHIRE REGIMENT (MIDLAND PIONEERS)

Instructions regarding War Diaries and Intelligence Summaries are contained in F. S. Regs., Part II. and the Staff Manual respectively. Title Pages will be prepared in manuscript.

Date	Hour	Summary of Events and Information	Remarks and references to Appendices
21.3.18		to the effect that all the Officers of his Company had become casualties & that he was in command of the Company. Six Officers & about 30 other ranks were sent up from N.Q. to reinforce "D" Coy	H.M.R. Capt
22.3.18	9.30am	Transport moved back to about H 13 c. Sheet 57c.	
		Two Platoons of C Coy withdrew in the morning to the Army Line about J 8 & 9 that 57c. & the remaining two in the afternoon.	
	6.0 pm	All Headquarter details moved up & dug in & occupied a line just behind the Army Line about J 14 b.	
	4.0 pm	Transport moved to PIONEER CAMP, LOGEAST WOOD (G1.6. Sheet 57c). 1 man of the transport was killed by shell fire.	H.M.R. Capt
		What remained of the Companies withdrew to the new line J14 b.	
		Total Casualties of the operation:— 2 Lieut. Hutchison, 2/Lt. C. Millward & 2/Lt. M. Baxter Killed in action Capt. R. Bentley, Lieut. H.H. Grundtvig MC, Lieut. H.J. Meggitt, 2/Lt. P.J. Wright, 2/Lt. O.H. Lowell 2/Lt. E. Bedson, 2/Lt. C.A.R. Thisson, Wounded Capt. J.C. Spencer, Lieut. A.L. Hicks, 2/Lt. M.H. Stevenson, & 2/Lt. A.J. Summers Missing 30 O.R. killed in action 106 O.R. Wounded 81. O.R. Missing	

2449 Wt. W14957/M90 750,000 1/16 J.E.C. & A. Forms/C.2118/12.

War diary for 11th Battalion, Leicestershire Regiment for 22 March 1918. (TNA:PRO WO 95/1601)

A few extracts from the diaries that contained secret or confidential information are now available in WO 154. Many of the maps, which were originally included in the diaries, have been extracted and added to WO 153.

The war diaries for the Wiltshire and Berkshire regiments have been transcribed and published at www.thewardrobe.org.uk.

Further reading

Ian FW Beckett, *The First World War: The Essential Guide to Sources in The National Archives* (PRO Publications, 2002)

Simon Fowler, *Tracing Your First World War Ancestors* (Countryside Books, 2003)

Norman Holding, *The Location of British Army Records, 1914–1918* (4th edition, Federation of Family History Societies, 1999)

— *More Sources of World War I Ancestry* (3rd edition, Federation of Family History Societies, 1998)

John Laffin, *Western Front Companion 1914–1918: A–Z Source to the Battles, Weapons, People, Places, Air Combat* (Sutton Publishing, 1997)

William Spencer, *Army Service Records of the First World War* (PRO Publications, 2001)

There are also a number of books on the experience of men in the trenches, including:

Max Arthur, *Forgotten Voices of the Great War* (Ebury Press, 2003)

Ilana Bet-el, *Conscripts: Lost Legions of the Great War* (Sutton, 1999)

Richard Holmes, *Tommy: the British Soldier on the Western Front* (Harper-Collins, 2004)

Richard van Emden, *The Trench* (Bantam, 2002)

— *Britain's Last Tommies: Final Memories from Soldiers of the 1914–1918 War, In Their Own Words* (Leo Cooper, 2005)

There are naturally a large number of websites devoted to the First World War (particularly for the Western Front), most of which are surprisingly good, although there is considerable overlap in subject matter. Among the best are:

www.1914-1918.net/menu_army.htm

www.bbc.co.uk/history/war/wwone/index.shtml

www.fylde.demon.co.uk/welcome.htm

www.firstworldwar.com

Chapter 16

THE SECOND WORLD WAR

Although historians have argued that service in the Army during the two world wars was broadly comparable, in fact there were many differences. Perhaps the main one is that most of the soldiers on the Western Front during the Great War saw some fighting at some stage. By the Second World War, however, technological advances meant that many more soldiers were involved in supporting the infantry than saw action themselves – what Churchill called the 'fluff and flummery behind the fighting troops'. It was estimated that 60,000 men were now needed to keep an infantry division of 20,000 in action in equipment, patrol and food.

British troops went to war by lorry or parachute, or advanced behind tanks, while communicating with their commanders by radio. New units were created either to use the technology, such as the Royal Armoured Corps (formed in 1939) and Parachute Regiment (1940) or to maintain it in the case of the Royal Electrical and Mechanical Engineers (1942). Other branches expanded to meet the new demands. The strength of the Royal Signal Corps, for example, was 34,000 in 1939. By 1945 it had risen nearly fivefold to 154,000.

The new technologies also ensured that the wounded received far better care than their fathers or grandfathers could have imagined. Over 178,000 sick and wounded men, for example, were evacuated by air from Burma to hospitals in India.

Deaths from the two great tropical scourges of malaria and dysentery were kept to a minimum. Soldiers fighting in the Burmese jungle were instructed to take daily doses of mepacrine and severe measures were taken against officers who did not enforce this. General Sir William Slim, the Commander-in-Chief, personally sacked three commanders who did not do so. Unfortunately the drug turned its takers yellow and there were the inevitable rumours about its effects on sexual potency. But it did its job: mortality from these diseases was less than ½ of 1 per cent. The official historians later concluded that: 'one important reason why the Allies were able to defeat the Japanese was that they were able to hold the malaria parasite in check, while the Japanese were not.'

For much of the time between 1939 and 1945, the experience of war for many men was not one of daring deeds at 'the sharp end', but rather of a sedentary existence in camps or depots across the country, filling in forms or supplying ammunition to troops in the front line.

Occasionally this caused resentment among the infantrymen, who bore the brunt of the fighting against the enemy. According to Bill Scully:

> *There was a vast difference between fighting troops and the support troops. Of course civilians didn't know. So you'd get somebody who'd never fired a shot in anger, but he could still say he was in the Eighth Army. Every time I came home, my neighbour would say 'My lad's in the Eighth Army. He's in Cairo. He's been there three years. I hope he's safe.' And I'd think, 'Blimey, wish I were.'*

The Army also no longer bore the brunt of casualties, as had happened a generation before. Proportionally the worst losses were in RAF Bomber Command, even so casualty levels during the invasion of north-west Europe in 1944–1945 at times reached the levels of the First World War.

Some 3,778,000 men (and an increasing proportion of women) served in the Army, three-quarters of whom were conscripts; 177,800 were killed, 239,600 wounded and 152,076 taken as prisoners of war.

This could lead to problems with recruitment, as many men preferred to enlist in the rather more romantic RAF or Navy. Not everybody, however, succeeded in joining the service of their choice, as one man remembered:

> *I didn't relish going into the Army, being well aware of the terrible suffering in the trenches. Within a couple of days of receiving my call up papers I was off to Uxbridge to volunteer for the RAF. The interviewing officers asked only two questions. One – could I type? I answered no. Two – could I do shorthand? No. 'Sorry,' he said, 'you'll have to go into the Army.'*

Conscription was introduced on the outbreak of war in September 1939. Eventually all men between eighteen and sixty were required to take some form of national service, although by no means everybody served in the armed forces. However, 1.5 million men volunteered without waiting for their call-up papers.

Joining the Army could be traumatic for many recruits, who were used to the comforts of home. Ernie Hamilton said:

> *Many young men were experiencing homesickness for the first time, and many tears flowed. Everything the Army issued seemed to be for the discomfort of the soldier. The blankets were very rough after the lovely sheets of home. Uppermost in my mind was the uniform, never in the right size, and the smell of the mysterious powder which we assumed was a delousing agent.*

There followed a period of basic training, which largely meant drill, to instil discipline, coupled with the seemingly endless polishing of equipment. Many recruits hated this aspect of their basic training. An Army Moral Report for May 1942 noted that 'the soldier particularly objects to "spit and polish" where its application serves no more apparent purpose.' It was said that while Russia bled, Britain blancoed.

Attempts to make the basic training, particularly the drill, more relevant were bitterly resisted by senior officers and NCOs. A committee found that 'A few senior Regular NCOs thought that the more pointless an order appeared to be and the more useless the task required, the greater the beneficial effect on discipline.' Sadly the reverse was true.

After about six weeks basic training, specialists ('tradesmen' as they were known) were separated out for further training in their specialities, such as wireless operating or driving. Indeed properly using the skills that a recruit brought with him proved difficult. Too often men were sent to trades or units for which they had no aptitude. A committee investigating the problem in 1941 found that less than half the engineers who had joined the Army had been mustered in an engineering trade.

It is little wonder that in the same year, the Adjutant General, who was responsible for the selection and training of recruits, wrote that: 'The British Army is wasting its manpower in this war almost as badly as it did in the last war, a man is posted to a Corps almost entirely on the demand of the moment. If we are to beat the Germans we must overhaul the whole system at once. The only way to obtain an efficient and contented army is to place the right man, as far as is humanely possible, in the right place.'

Psychological tests began to be introduced. The sociologist Tom Harisson found that the new procedure 'introduced an intelligent, intelligible system, making men feel valued for their private abilities and personal peculiarities.'

For the infantry, advanced training meant learning to handle weapons and to deal with battle situations. For many this was a deeply depressing experience. Brigadier CE Hudson wrote in the *Army Quarterly* in 1945 of:

'large classes of men seated in frozen apathy around an instructor, while he streamed forth the patiently learned patter. The soldiers either became NCOs themselves or contented themselves with acquiring the art of putting on the appearance of interested attention while remaining in fact entirely and hopelessly disinterested.'

Attempts were made to stimulate the men's interest through films and discussions. A number of training films were made, by the Army Directorate of Kinematography, which featured stars such as Thorold Dickinson, Carol Reed, Freddie Young, Peter Ustinov and Eric Ambler. However, the authorities were sceptical about how effective these films were. Lieutenant Colonel JW Gibb, of the Directorate of Military Training, grumbled: 'If an audience of trainees were presented with such films they would of course, as it was once aptly described by a general, "assume an art of unbuttoned ease" and probably go to sleep.'

Again, in contrast to the First World War, the relationship between officers and men was generally much more informal. It helped that most officers now came through the ranks, although they still tended to be better educated than the men they led. The system of selection was reformed in 1941 to ensure that more people from outside the traditional background were chosen. Even so, those selected to become officers often found the transition from the ranks daunting.

What was important to most men was the leadership they received from their officers and NCOs. Private Ian Kaye, of the Black Watch, remembered his platoon commander's speech before an attack on a German village in 1944: ' "OK? If we manage to get into the village without a fight, we move into clear the houses on each side of the road. Good luck and bloody good hunting." That was how he approached things. But he was a damned good bloke and he made you feel good too.'

Most soldiers were sent abroad during their service. For almost everybody, this was the first time they had ever left Britain and it was a novel experience. Tom Ridley was part of the 50th (Highland) Division who landed in Normandy on D-Day. He was a corporal in charge of six sappers (Royal Engineers). On leaving Portsmouth:

I went up on deck, took a long look at the white cliffs, and having decided in my opinion the chances of survival were about even, I said good-bye to my parents and my sister, and since there was no one else there I said it out loud. ... I tried to close my mind to the very unpleasant things that could happen to me any second now, and just put one foot in front of the other, hoping to get ashore without stepping into a hole, which was not advisable when carrying about 40 pounds of gear. As we made

*our way quickly across the beach the only dead man I recognised was a
lance-corporal from our platoon I'd visited Winchester with on our last day
of freedom.*

Most men, however, had their first sight of the wider world in happier
circumstances. Mel Lowry, of the Royal Tank Regiment, joined the First Army
in North Africa in early 1943:

*I remember smelling Algiers, this exotic spicy smell, before we saw it. And
then you saw it – all those glittering white buildings. We pulled in at the
docks and I saw the most ragged men I'd ever seen in my life – these were
the Algerian dockers.*

Despite all the training, nothing could prepare the recruits for their first taste of
the action. Nineteen-year-old Bill Scully, from Yorkshire, arrived in early 1944,
at Cerasola in southern Italy, during the night:

*In the morning you looked down and saw the litter and clots of congealed
blood, and only a few feet away you saw a dead German. He'd been shot in
the head. You could see the bullet hole plainly. Young man, about twenty,
blondish. I've seen him in my sleep many a time.*

The fighting deeply affected many men. It is thought that as many as one soldier
in ten suffered from some form of combat fatigue. As Colonel RH Ahrenfeld

British troops enter war damaged Naples in October 1943.

observed in the official history *Medical Services in War*: 'it is certain that modern mechanised warfare imposes on the individual a strain so great that men involved in active fighting, however basically stable they may be, will ultimately break down thus psychiatric casualties of the type described as "campaign exhaustion" are as inevitable as gunshot and shrapnel in modern warfare.' It was particularly a problem where men could not easily be evacuated back to Britain, such as in Burma and India: where there was also the fear that because of manpower shortages they might be sent to the front line again.

Even those who did not have this trauma felt an intense reaction to the prospect of going forward under fire. Alexander Baron, of the Wiltshire Regiment, spoke of the 'strange terror that afflicts the body regardless of his will, a twitching in his calves, a fluttering of the muscles in his cheeks, the breath like a block of expanding ice in his lungs, his stomach contracting, the sickness rising in his throat'. In the moments before a battle men prayed, fingered lucky charms or made weak jokes with fellow members of the platoon.

Whether soldiers were at the front line or working in camps and bases far from the fighting, entertainment was an important way to take people's minds off the war for a few hours. Army entertainment was coordinated by the Entertainments National Service Association (ENSA – which many said stood for 'Every Night Something Awful'). A stream of performers, good and bad, entertained the troops. After D-Day, ENSA had 4,000 acts on its books, and in a single month put on 13,500 stage shows and showed 20,000 films. Initially ENSA was only based in the United Kingdom, but by 1943 artistes were travelling the world. Perhaps the highlight was when, in 1944, Vera Lynn visited men of the 14th Army in Burma.

Not everyone was a willing member of the Army. Desertion was a considerable problem, partly because so many servicemen were stationed in Britain close to home and enduring years of tedious training or drill. Unlike in the First World War it was not punishable by death, although a number of commanders pressed the War Office to reintroduce it.

Desertion was a problem overseas. Over 10,000 men absented themselves in north-west Europe between June 1944 and victory in May 1945. The majority did so during the winter, when there was little fighting. In Italy, the Official History noted that: 'Throughout the campaign the rate of desertion and absence without leave was continually a source of anxiety.' Between September 1943 and June 1945, 5,700 men were convicted of desertion. The authorities concluded that the main reason was the stress on the fighting men, as time and time they were sent to the front without having had proper leave or the promise of their unit being sent back to Britain. Indeed leave was only granted to men serving in north-west Europe.

When the war was over, soldiers began to return to their loved ones glad to be alive, and thankful to be out of the Army, but still remembered their colleagues who didn't make it. For some it was difficult to adjust to civilian life again. As Joe Oldford, who served with the Canadian Cape Breton Regiment, said: 'You can't just come home after a war is over, take off your khaki, put on your civilian clothes and walk out as if nothing has happened.'

Despite the periods of boredom, the inappropriate training and the fear that combat brought, most soldiers enjoyed their time in the Army. There was a shared comradeship with others, the chance to see the world and perhaps a chance to develop new skills which often proved useful in 'civvie street'.

The returning veterans had a lot to be proud of. The British Army had made a major contribution to the defeat of the Axis powers. It had engaged in some memorable battles, and served with dignity despite the often terrible conditions and desperate circumstances. Donald Easton described the men who served with the 'Forgotten Army' – the 14th Army – in India and Burma:

> *He fought hand-to-hand battles practically every night, and his pals were shot down all around him. If he was wounded he had no hope of evacuation. Day after day he was promised relief which never came; and his own platoon, or section or just gang, got smaller and smaller. And he did all that on half a mug of liquid every 24 hours.*

The prevailing fear among soldiers was that they would be returning to unemployment as so many of their fathers had done. Fortunately a simple but fair system, however, ensured that men who had enlisted first or had skills in demand (particularly in the building trades) were demobilised first. Each man received clothing, including a suit, and the opportunity for training in skilled trades or grants to attend college or university. The booming post-war economy ensured that every man quickly found work.

16.1 Service records

Service records for both officers and other ranks are held by the Ministry of Defence's Army Personnel Centre, Historic Disclosures, Mailpoint 400, Kentigern House, 65 Brown St, Glasgow, G2 8EX, Tel: 0141-224 2023, email apc_historical_disclosures@btconnect.com. Veterans and next of kin can get copies of service papers for free; for people who are not next of kin a charge, currently £30, is made.

However, the Household Cavalry and Guards keep their own records:

Household Cavalry Museum, Combermere Barracks, St Leonards Road, Windsor, SL4 3DN, Tel: 01753 755112 www.householdcavalry.gvon.com/museum.htm.

Guards Museum, Wellington Barracks, Birdcage Walk, London, SW1E 6HQ

16.1.1 Officers

The vast majority of officers during the war were recruited from the ranks, although a few came from school or university Officer Training Corps (OTC). An officer had to spend at least a year in the ranks, before being recommended by his commanding officer and passing an interview and other tests at a selection board. Announcing this policy in 1939, the Secretary of State for War, Oliver Hore-Belisha, stressed that:

> *In this Army the star [that is the 'pip' on the lieutenant's uniform] is within every private soldier's reach. No one, however humble or exalted his birth, need be afraid that his military virtues will remain unrecognised. More importantly, no one, who wishes to serve in the Army need consider his state minimised by starting on the bottom rung.*

During the war some 200,000 commissions were issued to men, the vast majority of whom by 1945 had come via the ranks.

Although the Army was a much more democratic body than it had been in 1914, it still preferred men from the right public school background. The writer Alan Wood, had spent a year as a Gunner, before someone discovered that he had been educated at Oxford. 'I was promptly given a recommendation for a commission and a handsome apology for being kept in the ranks so long.'

Officers who came from the ranks often found it difficult to deal with the aristocratic and upper class elites who still formed the majority of pre-war officers or had entered from the OTC. Being an officer also involved some financial sacrifice, largely because junior officers' pay was less than that of an experienced NCO. There were also other expenses such as mess bills. Henry Longhurst, serving in the Royal Artillery, pointed this out in an April 1941 article in the *Sunday Express* headed 'No drinks or smokes on an officer's pay'. Matters improved when pay was increased and allowances were provided for the payment of mess expenses.

Commanding officers were also often reluctant to allow good non-commissioned officers to apply for commissions. The result was that fewer men were being promoted from the ranks than had been expected or met the Army's increasing needs.

The system was reformed in 1942 with the introduction of War Office Selection Boards, which thoroughly tested applicants for suitability regardless of rank or background. *Picture Post* declared that the new system was one of the most progressive initiatives of the war. Not everybody was so sure. George Macdonald Fraser felt that 'The general view throughout the Army was that [the boards] weren't fit to select bus conductors, let alone officers.' It was a view that Winston Churchill privately shared, although he was persuaded not to take any action.

Even so the proportion of officers who had been to public school fell from 84 per cent in 1939 to just 15 per cent at the end of the war.

Not withstanding their background, all officers are listed in the *Army Lists*, which were published quarterly during the war. For intelligence reasons the information given is not as comprehensive as would be found in peacetime lists. Even so it should be possible to track down promotions and the dates they were made. There were in fact two series – the Half Yearly and Quarterly, but both contain similar information. The National Archives has a complete set on the shelves in the Microfilm Reading Room, and copies can be found elsewhere.

Biographical details of officers in the Army Medical Corps are listed in Sir William Macarlane's *A List of Commissioned Medical Officers of the Army, 1660–1960* (Wellcome Library, 1968). A little bit more about researching medical personnel, including nurses, can be found at www.army-medical-history.co.uk.

16.2 Medals

16.2.1 Campaign medals

Serviceman and woman who met certain criteria (normally being present for ninety days in a particular theatre of operations) could claim up to five campaign medals.

Veterans could claim the medals when they were first issued in the late 1940s. Many did not do so then, but it is still possible to request them from the MOD Medal Office, Building 250, RAF Innsworth, Gloucester, GL3 1HW, Tel: 01452 712612 ext 8149 www.veteransagency.mod.uk/medals_folder/medals_campaign_medals.htm

Eight separate campaign medals were issued, although soldiers were only eligible for seven of them. As a cost-saving measure medals were not inscribed with the names of the individuals to whom they were awarded, although a number of individuals subsequently had inscriptions added privately.

The medals were:

1939–1945 Star	This was the basic war service star and was generally awarded to men who had completed six months' active service overseas. It was the only medal awarded to men who saw service in France and Norway in 1940 and Greece and Crete in 1941.
Africa Star	Awarded for one or more day's service anywhere in North Africa before 12 May 1943.
Burma Star	Awarded for service in India and Burma.
Italy Star	Awarded for service in Italy, the Balkans and southern France between 11 June 1943 and 8 May 1945.

Pacific Star	Also awarded to personnel who served in Hong Kong and Malaya.
Defence Medal	Awarded to all those who served in a military capacity in Britain, Malta and British colonies between September 1939 and May 1945, including civilians and members of the Home Guard.
War Service Medal	Issued to anyone who rendered twenty-eight days' service in uniform or in an accredited organisation.

More about these medals can be found in an article by John Sly in issue 35 of *Ancestors Magazine* (July 2005). They are also described in more detail in Peter Duckers, *British Campaign Medals 1914–2000* (Shire Publications, 2002), Robert W Gould, *British Campaign Medals: Waterloo to the Gulf* (Arms and Armour Press, 1994) and LL Gordon (et al), *British Battles and Medals* (Spink & Son Ltd., 1988).

16.2.2 Gallantry medals

As in previous wars details of all awards were published in the *London Gazette*, occasionally with a citation. At the very least you will get the man's name, service number, rank, regiment and the date when the award was made. These records are fully indexed and are available online at www.gazettes-online.co.uk. The National Archives has copies (in series ZJ 1) together with indexes in the Microfilm Reading Room and copies are often held by regimental museums and large reference libraries.

The award of gallantry medals to individuals may sometimes be mentioned in war diaries.

Some 181 awards for the VC were made during the war. There are several websites devoted to VC recipients, of which www. chapter-one.com/vc is undoubtedly the most comprehensive. It contains detailed biographies of all VC holders. A register of VC awards can be found in WO 98/8 at TNA, together with copies of citations and other information.

Most recommendations for the VC, DSO, MC, DCM and MM (and a very

The Africa Star – one of the seven campaign medals issued during the Second World War.

few Mentions in Despatches) are in series WO 373. To find them you will need to know when the award appeared in the *London Gazette*, and the theatre of war. A register, arranged by the date the awards were gazetted, is in WO 390/9-13. Extracts from the *London Gazette* for the award of the Military Cross (MC) are in WO 389.

The Distinguished Service Order (DSO) was normally only awarded to senior officers, while the Military Cross (MC) was awarded for acts of bravery to officers of the rank of captain or below. A register, arranged by the date the awards appeared in the *London Gazette*, is in WO 390/9-13. Extracts from the *London Gazette* for the award of Military Cross (MC) are in WO 389.

16.3 Casualties

16.3.1 Commonwealth War Graves Commission

The Debt of Honour Register (www.cwgc.org) will tell you where a man is buried, when he died, his rank and the unit he served with, and sometimes next of kin. The Commission will also provide the same service for people who write in or telephone their offices. Their address is: 2 Marlow Rd, Maidenhead, SL6 7DX, Tel: 01628-34221.

16.3.2 Army Roll of Honour

The Roll was compiled between the end of 1944 and March 1949 for the use of the War Office and the Commonwealth War Graves Commission. Individuals found in the roll died between 1 September 1939 and 31 December 1946, and also included are those deaths in service which were non-attributable (natural causes, etc.) as well as those, of course, who were killed in action or who died of wounds or disease. Apart from surname and forenames, Army service number and date of death, all other information (rank, first unit served in, unit serving in at time of death, place of birth, place of domicile and place of death), has been given numerical codes. The original is at The National Archives in series WO 304. It is also available on CD at many libraries and archives,

A unknown soldier of the Second World War.

and you can search the database online (by surname only) at www.military-genealogy.com.

16.3.4 Hospital records

Few, if any, records for individual men, either wounded or sick, survive. However, if the hospital to which your ancestor was admitted for treatment was in Britain (and you know which one) it might be worth checking the Wellcome Institute/National Archives database of hospital records, http://hospitalrecords.nationalarchives.gov.uk to see whether any patient records survive.

If you know at which hospital or casualty clearing station your man was a patient, you may find reports in series WO 222.

16.3.5 Prisoners of war

There are extensive collections of records at The National Archives which are described in a TNA Research Guide *British Prisoners of War 1939–1953*.

The IWM also has many documents including memoirs from former POWs and camp newspapers. The most comprehensive nominal listing of British and Commonwealth POWs is found in WO 392/1-26 at Kew.

Details of some 169,000 British and Commonwealth prisoners of all ranks held in Germany and German-occupied territories, with details of name, rank and service/Army number as well as regiment/corps, prisoner of war number and, presumably, their final camp location details can be found in *Prisoners of War: British Army 1939–1945* (JB Hayward & Son, 1990).

In the Research Enquiries Room at Kew there is a typescript research guide *British Prisoners of War, World War Two*, which includes full document references, dates and descriptions for reports on camps, nominal lists, escape and evasion reports, and miscellaneous reports.

TNA has approximately 140,000 Liberation Questionnaires (in series WO 344) completed by British and Commonwealth prisoners of war. They are arranged alphabetically by name with separate sections for those held by Germany and Japan. As well as giving personal details, name, rank, number, unit and home address, these records include details of the camps the men were in and their treatment by the enemy. Consequently, additional documentation is sometimes attached.

Escape reports completed by men who successfully escaped and made it back to Britain, are in three series WO 208/3298-3327, 4238–4276, 4368–4371. Every report has a narrative which describes an individual's experiences as an escaper, evader or prisoner of war. There is a card index to the first series of reports in the Research Enquiries Room at Kew. The others are arranged by surname of escaper.

Prisoners of the Japanese are listed in index cards in WO 345. The index cards are 56,000 pre-printed cards of uncertain provenance that appear to have been compiled by a central Japanese authority. The cards, with certain exceptions, record in Japanese and/or English, personal details and the camps the individuals were sent.

16.4 Other sources

16.4.1 War diaries

While on active service, Army headquarters, formations and units were required to keep war diaries recording their daily activities. They generally consist of war diary or intelligence summary sheets located at the beginning of each diary which record the date of each entry, the unit's location, a summary of events and any remarks or references to appendices. The appendices, which make up the larger part of each diary, may include strength and casualty returns, routine orders and administrative instructions, operation orders and instructions, reports, messages, location statements, intelligence summaries, and occasionally photographs, sketches, maps and traces. They can be of particular interest to family historians as they may contain unit orders and lists of men being transferred in or out of the unit.

The war diaries vary greatly in the amount and level of detailed information they contain. Their quality generally reflects the skill, dedication and enthusiasm of the officers in charge of compiling them. They are a historical record of a

War diary for 1st Battalion, Dorset Regiment for D–Day 1944. (TNA:PRO WO 171/1284)

unit's administration, operations and activities and rarely record information about individual men.

It is unusual for individuals, particularly ordinary soldiers, to be mentioned by name. However, the lack of personal names is not necessarily important. Knowing that your ancestor was in that unit is sufficient to build up a picture of what he was engaged in.

The war diaries are arranged by theatre of operations, so you need to know where your ancestor served. If you are not sure, you will need to consult the Orders of Battle (sometimes called ORBATs) in series WO 212 at TNA, although it is probably easier to use one of the online Orders of Battle. Probably the best one, although strictly the information isn't taken from orders of battle, is at www.regiments.org – the site is easy to use and arranged by regiment rather than by campaign or other odd ways devised by the webmaster. The best source in print is Lieutenant Colonel HF Joslen, *Orders of Battle: United Kingdom and Colonial and Dominion Formations in the Second World War* (HMSO, 1960, reprinted by The Naval & Military Press, 2003).

The references are:

Description	Series	Coverage
War Office Directorates	WO 165	
Home Forces	WO 166	Forces stationed in the United Kingdom
British Expeditionary Force	WO 167	France 1939–1940
North-West Expeditionary Force	WO 168	Norway 1940
Middle East Forces	WO 169	Egypt and Libya, East Africa, Iraq, Iran
Central Mediterranean Forces	WO 170	Italy, Greece and Austria
North-West Europe	WO 171	France and Germany, 1944–45
South-East Asia Command	WO 172	India and Burma
West Africa Forces	WO 173	
Madagascar	WO 174	
British North Africa Forces	WO 175	Algeria and Tunisia 1942–1943
Various smaller theatres	WO 176	Mainly Caribbean, Faeroes, Fiji, Gibraltar, Iceland
Medical services	WO 177	
Military Missions	WO 178	
Dominion Forces	WO 179	
GHQ Liaison Regiment	WO 215	
Special Services	WO 218	

16.4.1 Courts Martial

A court martial is a court convened to try an offence against military discipline, or against the ordinary law, committed by a person in one of the armed services.

Registers of courts martial are mainly in WO 213, with surviving case papers in WO 71 mainly for the most important trials. There are some miscellaneous records in WO 93, including details of death penalties carried out after 1941. Some records relating to individual cases of court martial are closed for seventy-five years from the last date on each file, which means that many files relating to the Second World War are still not available. More information is given in a TNA research guide *British Army: Courts Martial, 17th–20th Centuries*.

Further reading

Simon Fowler, *Tracing your Second World War Ancestors* (Countryside Books, 2006)

Phil Tomaselli, *Second World War 1939–1945* (Federation of Family History Societies, 2006)

There are a number of online Reader's Guides produced by The National Archives which explain these records in more detail.

Thousands of books have been published about the Second World War. Here is a selection of recent introductory volumes:

Max Arthur, *Forgotten Voices of the Second World War: A New History of the Second World War and the Men and Women Who Were There* (Ebury Press, 2005)

Peter Calvocoressi (ed), *The Penguin Book of the Second World War* (Penguin, 2001)

Ian Dear (ed), *The Oxford Companion to the Second World War* (Oxford UP, 2001)

John Ellis, *The World War II Databook* (Aurum Press, 1993)

Martin Gilbert, *Second World War* (Weidenfeld and Nicolson, 1989)

Richard Holmes, *Battlefields of the Second World War* (BBC Books, 2001)

John Keegan, *Second World War* (Pimlico, 1997)

Chapter 17

THE ARMY IN PEACETIME, 1919–1969

D uring the fifty years between 1919 and 1969 when Britain was formally at peace, the Army was still engaged in military actions of one kind or another. Indeed, 1968 was the only year since the end of the Second World War in which no soldier was killed in action.

The Army and the officers and men who served in it had very different experiences between 1919 and 1939 and after 1945. This was largely due to public opinion and the perceived threats to British interests in the wider world.

During the interwar period the Army reached the nadir of its popularity with the politicians and the public. It was inevitable and understandable that society of the period should react against the losses of the First World War. In 1919 it looked to the planners as if there were no enemies on the horizon. Britain's enemies (particularly Germany) had been comprehensively defeated or, in the case of the Soviet Union, were not in a position to attack.

As a consequence ministers starved the armed forces of funds. Priority was given to the needs of the Royal Air Force and Royal Navy, because it was believed that they would deliver victory more efficiently than the Army and certainly for less loss of life. Even within the Army emphasis was placed on mobility and resources were directed towards tanks and light armoured cars. Its early lead in designing tanks and armoured vehicles (the tank had been a British invention in 1916) was lost. When rearmament belatedly increased the Army's resources it was too late to make up lost ground before the war — indeed, British tank design only began to catch up in the middle of the Second World War.

By the end of 1921 the Army had returned to its pre-war size, with just 217,500 officers and men on its books. Regiments, including the five which had been based in southern Ireland, were disbanded. Particularly hard hit were cavalry regiments with many mergers.

Even in periods of economic depression recruitment proved difficult, despite improvements to pay and conditions. Young men found that both the RAF and the Navy seemed more modern and more romantic, with less chance of being cannon-fodder should another world war break out.

Even so emphasis was increasingly placed on technology. New branches were established for Signals, Ordnance and most obviously of all the Royal Tank Corps, which was formed in 1923. Attempts were made to recruit skilled workers. Better training was offered to new recruits and half hearted efforts were made to instil initiative and independent thinking among ordinary soldiers. And, perhaps most significant of all, a scheme was introduced to allow men from the other ranks to more easily become officers, by offering places at the Royal Military College at Sandhurst to bright soldiers. Even so as late as 1937, the Army was short of 980 officers and 20,000 soldiers. This shortage only began to be rectified with the introduction of National Service a few months before the outbreak of war in 1939.

As had long been the case, the Army remained largely based in India and the colonies, notably the newly acquired mandated territories of Palestine and Iraq (which then, as now, required pacification). For twenty years it was stationed away from Europe by repeated Cabinet affirmations that no expeditionary force would again be sent to the Continent. Even when military planning for a possible war began to be taken seriously, it was assumed, as in 1914, that one division would be sent to aid the French, because it was believed the Maginot line which protected France from Germany was impregnable.

It is little wonder that as late as November 1938, Prime Minister Neville Chamberlain is supposed to have remarked that the British Army was so small that it was hardly worth worrying about.

The world of 1945 was very different from that of 1939. The immediate tasks before the military commanders were to maintain preparedness for war with Russia, and to disengage from the Empire. Within three years of the defeat of Japan, Britain had left India and Palestine and was beginning to fight Chinese communists in the jungles of Malaya. In 1950 British troops helped defend South Korea from aggression from the Russian supported North.

Perhaps the greatest change was the slow realisation that the United Kingdom was no longer a great power. The war had impoverished the nation and Britain could not afford a worldwide military presence. Defence reviews in 1957 and 1967 made clear the need to cut back our military commitments, particularly 'East of Suez' as the phrase went. The Sandys Review of 1957, for example, cut the Army from 373,000 to 165,000, although this figure soon rose to 195,000 to meet all commitments. In 1951, there were eighty-five infantry battalions, by the early 1960s this had been reduced to sixty. The Royal Armoured Corps lost seven of its thirty regiments and the Royal Artillery twenty regiments plus another fourteen when responsibility for anti-aircraft defence was passed to the RAF. All this at a time when the Army was fully committed around the world.

It was increasingly obvious that our future lay in Europe as a close ally of the United States. The formation of the North Atlantic Treaty Organisation (NATO) in 1949 formalised this. Indeed when the Americans disapproved, as they did over the Suez crisis of 1956, it was very clear that the British could no longer act as a truly independent power.

The Army had to respond as best it could. Since the eighteenth century one constant problem had been the difficulty in recruiting men. Full employment until the mid-1970s meant that the service had increasingly to compete with the private sector. Wages and conditions continued to improve and more emphasis was placed on training, both to maintain the increasingly involved kit, but also with a view of enabling the serviceman to find a job once he had been discharged.

To an extent this was disguised during the period of National Service between 1949 and 1963, which brought hundreds of thousands of young men (willing or otherwise) into the Army. For most their service was a waste, but some saw action in Korea, Aden and other international hotspots. And, if there were fewer soldiers at the 'sharp end', they remained the most important and most visual part of the Army, whether engaged in peacekeeping operations in Kenya, Cyprus or, from 1969, Northern Ireland, or engaged in training exercises in Germany or Britain. Many of their operations were 'low intensity' or peacekeeping which would have been familiar to their fathers or grandfathers. And the soldierly virtues remained much the same; the need to be skilful in the use of weapons and the use of surroundings, the requirement to be alert, steadfast and brave and the ability to endure hardships. But above all there remained the loyalty to comrades and to the regiment.

17.1 Service records

Service records for men who left the Army after December 1920 are with the Ministry of Defence, Army Personnel Centre, Historic Disclosures, Mailpoint 400, Kentigern House, 65 Brown St, Glasgow, G2 8EX, Tel: 0141-224 2023, email apc_historical_disclosures@btconnect.com. Veterans and next of kin can get copies of service papers for free. A charge, currently £30, is made for all others. It is a good idea to search the pages on the Veterans Agency website www.veteransagency.mod.uk before you request a service record.

The Household Cavalry and Guards regiments keep their own records:

- Household Cavalry Museum, Combermere Barracks, St Leonards Road, Windsor, SL4 3DN, Tel: 01753 755112, www.householdcavalry.gvon.com/museum.htm
- Guards Museum, Wellington Barracks, Birdcage Walk, London, SW1E 6HQ

It is, of course, possible to build up brief details of the career of an officer from his entries in the published *Army Lists*. The National Archives (TNA) has a complete set on the open shelves in the Microfilm Reading Room. Regimental and military museums should also have sets.

By the 1920s most regiments produced magazines and journals which commented on life in the units and often mentioned individual soldiers and, especially, officers and senior NCOs. The post-1945 period also saw the publication of official newspapers for the troops, which will sometimes mention individuals but more importantly give a flavour of what life was like. Again regimental museums should have sets, particularly of journals. The British Newspaper Library should have copies of the newspapers and many of the journals. For more information see Chapter 1.

17.1.2 National Service

Between 1949 and 1963 tens of thousands of young men served a two year period (initially eighteen months) with a period of three years and six months (four years) in the reserves. Service records are still with the Ministry of Defence, although you may be able to get a flavour of an individual's experiences from the Unit Historical Reports in WO 305.

The Imperial War Museum and National Army Museum, as well as regimental museums, have many memoirs and artefacts relating to National Service (indeed the National Army Museum (NAM) recently mounted an exhibition on the subject). There are also several books, including: Trevor Royle, *The Best Years of their Lives* (John Murray, 1986), Keith Miller, *730 Days to Demob: National Service and the post-war British Army* (National Army Museum, 2003), and Tom Hickman, *The Call up: a History of National Service* (Headline, 2004).

There are several websites devoted to the subject, largely consisting of reminiscences, including: http://maxpages.com/lionelbeck/National_Service and www.nationalservicememoirs.co.uk.

17.2 Medals

Medal rolls for those issued for participating in an interwar campaign are in WO 100 at The National Archives. Rolls for post-war campaign medals are still with the MOD Medal Office, Building 250, RAF Innsworth, Gloucester, GL3 1HW, Tel: 01452 712612 ext 8149 www.veteransagency.mod.uk/medals_folder/medals_campaign_medals.htm.

The most common campaign medal was the Military General Service Medal, for which clasps were awarded for service in some sixteen small wars between 1918 and 1962 in Iraq, Palestine, India, the Suez Crisis of 1956, Malaya and Arabia. It was replaced by a new General Service Medal in 1962, with clasps for service in Northern Ireland, Borneo and other places where British forces served. There have been separate campaign medals for the Korean War, the South Atlantic (Falklands War), and the Gulf Wars. Recently it was decided to retrospectively issue a campaign medal to men who were around the Suez Canal between 1951 and 1954. The award of these medals is summarised in Peter Duckers, *British Campaign Medals 1914–2000*(Shire Publications, 2001).

The awards of gallantry medals are listed in the *London Gazette*, which can be found online at www.gaztettes-online.co.uk.

17.3 Casualty records

The names of men who lost their lives on active service are recorded in the Commonwealth War Graves Commission's Debt of Honour Register. The Register is online at www.cwgc.org or you can write to their headquarters and they will send you details: 2 Marlow Rd, Maidenhead, SL6 7DX, Tel: 01628-34221.

British losses during the Korean War (1950–1953) were 793 killed and 2,878 wounded. Deaths are listed online at www.uk.or.kr/wargrave (with pictures of many of the graves) and in *Casualties sustained by the British Army in the Korean War, 1950–53: Compiled from Lists Published in The Times Newspaper* (Haywards Heath, undated).

17.4 Campaign records

17.4.1 Inter-war

Few records, if any, survive at The National Archives for individual units engaged in colonial campaigns of the period, which included Iraq (Mesopotamia), Palestine and India. However there are policy and other files in the War Office (WO), Colonial Office (CO) and Cabinet Office (CAB) records which may shed light. Using TNA's online catalogue should turn up references.

For India, including the 3rd Afghan War of 1920, it is worth trying the Oriental and India Office Collections at the British Library. The Imperial War Museum, regimental museums and, to a lesser degree, the National Army Museum may also have records.

17.4.2 Post-war (1945–1969)

It is well known that, with the exception of the year 1968, British forces have been engaged continuously in one action or another since 1945. The geographical range and variety of military tasks involved during these operations has been immense. The major campaigns in Korea, Suez and the Falklands can be set against peacekeeping duties in Bosnia, Sierra Leone and Northern Ireland. No western country has had as much experience (or as much success) as Britain in counter-insurgency as diverse as Palestine 1945 to 1948, the Mau Mau troubles in Kenya or the EOKA campaign in Cyprus.

Records for a few campaigns can be found at The National Archives. There are restrictions however. There are few records after 1965 and those of campaigns, such as Borneo, Malaya and Northern Ireland, which are still regarded as being controversial, have yet to be opened.

Between 1946 and 1950, Army units, wherever they were, compiled Quarterly Historical Reports. They are similar in format to war diaries, but are not as detailed. The references are:

Area	Class
British Army of the Rhine	WO 267
British Element Trieste Force	WO 264
British Troops Austria	WO 263
Caribbean	WO 270
Central Mediterranean Forces	WO 262
East and West Africa	WO 269
Far East	WO 268
Gibraltar	WO 266
Home Forces	WO 271
Malta	WO 265
Middle East (including Palestine)	WO 261

Quarterly Historical Reports were replaced by the Unit Historical Records in 1950; these are in WO 305. The list has a subject key. Operation Record Books for the Army Air Corps between 1957 and 1969 are in WO 295.

There are also more detailed operational records for particular campaigns:

Palestine 1945–1948	War diaries and headquarters papers	WO 191
	6 Airborne Division Papers	WO 281
Korea 1950–1953	War diaries	WO 281
	Historical records and reports	WO 308
	UN Command Operations papers	DEFE 12
Suez 1956	Headquarters papers and war diaries	WO 288
	Maps	WO 322
Oman 1957–1961	HQ British Forces Gulf Area	WO 337
South Atlantic 1982	Surrender documents	DEFE 14

The National Archives has numerous records relating to British and Commonwealth prisoners captured by the North Koreans. The papers of the Directorate

The Imperial War Museum has major collections of material relating to wars of the twentieth century.

of Military Intelligence contain lists of the British and Commonwealth personnel who were known, or believed, to be prisoners in Korea between January 1951 and July 1953. These lists are in WO 208/3999. The Historical Records and Reports on the Korean War in WO 308/54 also contain a list of Commonwealth prisoners, compiled in January 1954. Correspondence with returned POWs, and on personnel missing or presumed dead, is in WO 162/208-264 and DO 35/5853-5863.

The Imperial War Museum, the National Army Museum and regimental museums all have collections relating to the British Army's post-war activities.

There is an excellent website devoted to the period, www.britains-smallwars.com with histories, anecdotes from the men who were there, a detailed bibliography, and links to other sites on almost every aspect of Britain's post-war Army.

Further reading
There are, of course, numerous books on individual campaigns. A good overview is John Pimlott's *British Military Operations 1945–1984* (Hamlyn, 1984). Another interesting book is Robin Neillands, *A Fighting Retreat: the British Empire 1945–1997* (Hodder and Stoughton, 1996). The stories of some of the men who were there are told in Charles Allen, *The Savage Wars of Peace* (Michael Joseph, 1990).

Appendix 1

A SHORT GUIDE TO ARMY SERVICE NUMBERS

Until numbers were introduced men were generally identified by name, by trade, by profession, and by place of birth. Clearly, where there were several men of the same name in a particular unit, opportunity for confusion was rife, but even then, in general, only if those individuals served in the same company, were they separately identified in the pay lists.

In fact it was not until November 1829 that the War Office issued instructions to the effect that every soldier in the infantry and cavalry should have a unique regimental number. The method was to give the numbers in accordance with length of service. Part of the instructions included a name which was to pass into British folklore:

> *Every soldier is to communicate to his friends the number by which he is known in the Regiment, and to acquaint them that in all inquiries which they make after him ... they are to state such number; as 'Thomas Atkins, 5th Foot, No.55'.*

Originally numbers were given sequentially to soldiers as they joined the regiment. By the time of the Crimean War the number sequence in many infantry regiments had reached the high 3,000s or even 4,000s, but in 1857, the number sequence for the infantry was stopped and re-started with numbers back to 1.

This means that, theoretically, a situation could arise where two men serving in the regiment at the same time had the same number, one dating from the period circa 1830–1857, the other from 1857 onwards.

The next change in the number system came in 1873 with the first phase of the Cardwell reforms, with the introduction of brigades linking two regiments

A selection of regimental badges.

or battalions together. The system changed again in 1882 after the establishment of two battalion regiments and linked Territorial and Militia battalions.

This new system may have seemed simple and effective, but unfortunately there was still confusion in numbering, because men were not given new numbers – the three numbering systems continued. However, if a soldier changed his status for any reason (such as changing regiment, or rejoining for an extra period of service) he lost his pre-1881 number and was given a new one. This can cause tremendous confusion for researchers, as a soldier who enlisted under one number could be discharged with a completely different one. The change of number is often recorded only in the muster rolls in WO 16 (up to 1898), if no service record has survived.

This system ran unchanged from 1882 until 1920, although because of the huge expansion of the Army during the First World War, many units introduced variants to meet local needs. Again a man's number changed if he changed regiment, which was increasingly common in the last eighteen months of the war. Although all the service numbers should appear on an individual's medal index card, this is by no means always the case.

It was not until 1920 that Army numbers, as opposed to regimental numbers, were introduced, which were allocated to a soldier on enlistment and which he kept all his career, no matter how many times he changed his unit. The man who was assigned Army number 1 was Regimental Sergeant Major George James Redman of the Royal Army Service Corps. He had enlisted in the Corps aged fourteen in 1888, later seeing service in South Africa and the First World War. He was discharged shortly after receiving his new number.

More about the new system can be found at www.airbornerecce.com/dtroop/tables/anumbers.htm.

The best introduction to the subject is John Sly, 'Army Numbers to 1920' *Ancestors Magazine* (23) July 2004, upon which this appendix is based.

Appendix 2

PROBLEM SOLVING

Because of the variety of records which survives at The National Archives and elsewhere, tracing Army ancestors is not as difficult as many other forms of genealogical research. Even so there are pitfalls – such as interpreting the records correctly – and certain records may be missing for the period or individual you are interested in.

However problems may arise – two of the most common are: 'how do I identify the unit a man served with', and 'the records are missing, what alternatives are there?'

How to identify a regiment

To trace an ordinary soldier you need to know the unit in which he served, as records are generally arranged by regiment, particularly before the 1880s. Even later there are likely to be several individuals with the same name, even within the same regiment, so you are going to need to find out which one is your ancestor.

If you don't have this information here are some hints.

- If you have access to the internet it is worth entering the name of the man you are searching for into Google, or another search engine, to see what emerges. An alternative is to subscribe to one of the online mailing lists devoted to military genealogy or, perhaps, a particular surname to see whether a member can help you. They are described in Chapter 1.
- If you know where a man enlisted or lived it may be worth checking to see which units were based locally at the time you think he joined up. There are several ways to find out. The *Army Lists* (see Chapter 4) will tell you where regiments were based.

In addition, The National Archives has Army Monthly Returns between 1859 and 1950 in series WO 73, which show the distribution of the Army month by month, by divisions and stations and by regiments in numerical order. They give the station of each battalion or company, the numbers of officers and rank and file present or absent, and other statistical information. Earlier records, from 1754, are in WO 17. More detailed records (known as orders of battle) for the First World War are in series WO 95 and for the Second World War in WO 212 at Kew.

- Regimental muster rolls (see Chapter 5) may tell you where recruiting parties were based. These records were the basis for John M Kitzmiller's *In Search of the 'Forlorn Hope': a Comprehensive Guide to Locating British regiments and their Records (1640–WWI)* (3 vols., Manuscript Publishing Foundation, 1988). This reference book is widely available despite being deeply flawed. A critique of Kitzmiller's book can found at www.regiments.org/about/faq/kitzmill.htm.

- If your man was in the Army on a census night between 1841 and 1901 and based in Great Britain (not Ireland) then you should be able to pick him up in the census. Census records are all online with detailed indexes, so it should be easy to pick him up. (See Chapter 1 for more details.)

- The registers of births of children of Army personnel at the Family Records Centre are indexed and it may be possible to determine a regiment from them, if you have some idea of when children were born or the area where a soldier served. (These registers are described in Chapter 1.)

- If you know where your ancestor was living between 1842 and 1862 for England or Scotland (1842 and 1882 for Ireland and overseas) you may be able to pinpoint the regiment from the records of payment of pensions in series WO 22 and PMG 8 at The National Archives, which include the names of regiments in which individuals served. (These records are described in Chapter 9.)

- And, if all else fails, local newspapers may for the period may be able to help, recording the stationing of regiments and possibly the passage of recruiting parties. If the man lived to a great age, fought at a notable battle or had a distinguished career there may also be an obituary. (See Chapter 1.)

Identifying uniforms and badges

It can help considerably if you have a photograph of your ancestor in uniform. As early as the 1850s soldiers were having photographs taken of themselves to give to family and friends, so they turn up fairly often in family papers, particularly for the First World War. The uniforms they were wearing may give clues about the unit they were serving with and even the soldier's career in the Army.

The Royal Horse Artillery at Tel-el-Kebir, 1885.

With some effort and a bit of luck you should be able to find out more about the man and his service from looking carefully at the picture. The first thing to do is to try to identify his regiment from a cap badge or belt buckle. Chevrons on the sleeves of his uniform would indicate that he was a non-commissioned officer, while other patches might denote the number of times he was wounded, qualifications for riflemanship or other skills, and during the two world wars the division to which his unit belonged. Over his heart should be the medal ribbons, or possibly the medals themselves. Pips on his shoulder and a smarter cut of uniform would suggest that the sitter is an officer.

However, with the bewildering number of badges, stripes and other insignia it is easy to get confused. Fortunately there are a number of guides to help you identify military uniforms. One of the best (and simplest) is Iain Swinnerton, *Identifying your World War I Soldier from Badges and Photographs* (Federation of Family History Societies, 2001). Even though it refers to the First World War, most badges had been introduced well before 1914 and continued to be used up to and after the Second World War. Jon Mills, *From Scarlet to Khaki: Understanding the Twentieth Century British Army Uniforms in your Family Album* (Wardens Publishing, 2005) is an excellent guide to interpreting the other badges which might appear on an Army uniform apart from unit badges.

There are several websites devoted to military badges which may help, although they are not always complete and the quality of the reproduction of the badges themselves can vary. One of the better ones is www.militarybadges.org. uk/badges/badgestart.htm. The site also has many pages of photographs of old soldiers, with their life stories where known. Incidentally, the site's owner, Roger Capewell, has also self-published a guide to the subject called *Military Badges for Collectors and Historians*. Full details can be found at www.militarybadges. org.uk/badges/book_advert.htm. Another useful site, particularly for the First World War, is www.geocities.com/Athens/Acropolis/2354/army.html.

David J Barnes has written a number of articles, including 'Identification & Dating Military Uniforms' which appeared in Don Steel and Lawrence Taylor (eds), *Family History in Focus*, (Lutterworth Press, 1984). See also his 'A Brief Guide to the Identification and Dating of British Military Photographs' in *Family History Monthly* (no. 65) and 'Identification and Dating British Military Uniforms' in *Family and Local History Handbook* (6th Edition, Genealogical Services Directory, 2002).

Mr Barnes will also identify military uniforms for a small fee (currently £5 per photograph). He can be contacted at 148 Parkinson Street, Burnley, BB11 3LL or visit www.rfc-rnas-raf-register.org.uk.

My ancestor died on active service, how do I find out more?

- Provided you know approximately when a man died on active service and he was in the Army after 1914, he will appear in the Commonwealth War Graves Commission's Debt of Honour Register (www.cwgc.org), which will give you his regiment, regimental number and, of course, when he died and where he is buried.
- This facility is not available before 1914, although if you know in which campaign he died during the nineteenth century then the casualty rolls (Chapter 7) may help.
- If the soldier died in service, another possibility would be to check the records of soldiers' effects, which survive between 1810–1822, 1830–1844 and 1862–1881. They are in WO 25, arranged by initial letter of surname, and they give the regiment. However, these records are unlikely to be of use if the soldier died owing money to the Army. Later records are with the National Army Museum (Chapter 7).

His First World War service record does not survive, what alternatives are there?

- You can find very brief details from his Medal Index Card. They are at Kew or you can check them online at www.documentsonline.nationalarchives. gov.uk.
- War diaries for his unit will give you an idea of what he did day by day. However, it is unusual for other ranks to be named.

- If he was killed then you need to check the Commonwealth War Graves Commission website (www.cwgc.org) and the Soldiers Died in the Great War database (www.military-genealogy.co.uk) (Chapter 15)
- If he served after 1920 or during the Second World War his record will still be with the Ministry of Defence (Chapters 16–17)

My ancestor was in the Royal Flying Corps, but I can't find anything about him.

The Royal Flying Corps (RFC) was established in 1912 as the Army's flying arm and absorbed into the new Royal Air Force (RAF) on 1 April 1918. Men in the Corps were then given a chance to transfer to the RAF or remain in the Army. Service records for those who transferred and left the service before the end of 1920 are in AIR 76 for officers and AIR 79 for other ranks. Otherwise you will need to contact the Ministry of Defence, P MAN 2b(1), RAF PMC HQ PTC, RAF Innsworth, Gloucester GL3 1EZ.

Records of squadrons and other material can be found in series AIR 1 at Kew.

More details are given in a TNA Research Guide *Royal Air Force (RAF), RFC & RNAS: First World War, 1914–1918: Service Records* or in William Spencer's *Air Force Records for Family Historians* (PRO, 2000).

Appendix 3

TNA RESEARCH GUIDES ON TRACING SOLDIER ANCESTORS

The National Archives publishes an excellent series of Research Guides which can be downloaded from its website at www.nationalarchives. gov.uk/catalogue picked up in the reading rooms or you can telephone 020-8392 5200 for copies. They explain in simple terms the records and how they can help you in your research. The list at time of writing is:

Armed Forces: Sources for the History of
Auxiliary Army Forces: Volunteers. Yeomanry, Territorials & Home Guard 1769–1914
British Army Lists
Campaign Records 1660–1714
Campaign Records 1714–1815
Campaign Records 1816–1913
Campaign Records 1914–1918, First World War – War Diaries
Campaign Records 1939–1945
Campaign Records 1945 onwards
Courts Martial: Seventeenth – Twentieth Centuries
Courts Martial: First World War 1914–1918
Muster Rolls and Pay Lists c1730–1898
Officers' Commissions
Officers' Records 1660–1913
Officers' Records 1914–1918
Soldiers' Discharge Papers 1760–1913
Soldiers' Papers 1914–1918

Reading a war diary in the Document Reading Room at The National Archives, Kew.

Soldiers' Pensions 1702–1913
Useful Sources for Tracing Soldiers
First World War 1914–1918: The Conduct of the War
Conscientious Objectors and Exemptions from Service
Disability and Descendants' Pensions
Intelligence Records in The National Archives
Maps and Plans in The National Archives: Overseas Relations
Maps in The National Archives
British Armed Services: Campaign and other Service Medals

British Armed Services: Gallantry Medals
British Armed Services: Gallantry Medals. Further Information
Civilian Gallantry Medals
Medieval and Early Modern Soldiers
Military Maps of the First World War 1914–1918
Military Maps of the Second World War
Militia 1757–1914
Nurses and Nursing Services: British Army
Prisoners of War (British) c1760–1919
Prisoners of War (British) 1939 –1953
Probate Records
Service Records Second World War
Tudor and Stuart: Militia Muster Rolls
War Dead: First and Second World Wars
Wills and Death Duty Records after 1858
Wills before 1858: where to start
Women's Military Services: First World War

Appendix 4

WAR OFFICE RECORD SERIES AVAILABLE ONLY ON MICROFILM

WO 25   Miscellaneous registers 1660–1938

WO 36   Military Headquarters, North America: Entry Books, American Revolution 1773–1799

WO 42   Officers' Birth Certificates, Wills and Personal Papers 1755–1908

WO 65   Printed Annual Army Lists 1754–1879

WO 76   Records of Officers' Services 1764–1961

WO 97   Soldiers' Documents 1760–1913

WO 100   Medal rolls 1793–1949

WO 208   Directorate of Military Operations and Intelligence, Files 1917–1974

WO 228   Allied Forces, Mediterranean Theatre: Military Headquarters Papers, Second World War 1943–1946

WO 229   Supreme Headquarters Allied Expeditionary Force and 21 Army Group: Microfilms 1943–1945

WO 329   Service Medal and Award Rolls, First World War 1917–1926

WO 338   Officers' Services, First World War, Index to Long Number Papers 1870–1922

WO 343   South East Asia Command British Army Aid Group, China: Microfiche Copies 1942–1945

WO 356   Judge Advocate General's Office, Military Deputy's Department: War Crimes, South East Asia, Card Indexes, Second World War 1940–1957

WO 101 Meritorious Service Awards, Registers 1846–1919

WO 102 Long Service and Good Conduct Awards, Registers 1831–1953

WO 116 Royal Hospital, Chelsea: Disability and Royal Artillery Out-Pensions, Admission Books 1715–1913

WO 117 Royal Hospital Chelsea: Length of Service Pensions, Admission Books 1823–1920

WO 120 Royal Hospital, Chelsea: Regimental Registers of Pensioners c1715–1857

WO 121 Royal Hospital, Chelsea: Discharge Documents of Pensioners 1782–1887

WO 137 Earl of Derby, Secretary of State for War: Private Office Papers 1921–1923

WO 140 School of Musketry, Hythe: Records of Experiments and Trials 1853–1928

WO 144 Inter-Allied Armistice Commission: War Diary, and Despatches of Chief of British Delegation 1918–1920

WO 146 Distinguished Conduct Medal, Submissions to Sovereign 1855–1909

WO 357 South East Asia Command, War Crimes Branch: Record Cards 1945–1948

WO 363 Soldiers' Documents, First World War 'Burnt Documents' 1914–1920

WO 364 Soldiers' Documents from Pension Claims, First World War 1914–1920

WO 372 Service Medal and Award Rolls Index, First World War c1914–1922

WO 373 Recommendations for Honours and Awards for Gallant and Distinguished Service (Army) 1935–1990

WO 387 Order of the British Empire Register 1920–1949

WO 388 Exchange of Army Decorations between Britain and the Allies Registers 1914–1928

WO 389 Distinguished Service Order and Military Cross Registers 1911–1982

WO 390 Distinguished Service Order Register 1886–1945

WO 391 Distinguished Conduct Medal Register 1854–1920

In addition there are collections of other material in the Microfilm Reading Room which might be of use:

GRO indexes*
International Genealogical Index (IGI)*
Index to Prerogative Court of Canterbury (PCC) wills 1750–1857*

London Gazettes (ZJ 1)**
National Probate Calendar 1857–1943
Parliamentary papers*
Soldiers Died in the Great War++

* Also available online
** Partly available online
+ index available online
++ also available on CD in TNA Library and online

Appendix 5

ARMY RANKS

A rmy ranks have changed surprisingly little over the centuries and remain largely as they were 300 years ago.

The basic infantry commissioned ranks, at regimental level, were as follows (in descending order):

Lieutenant Colonel: the Commanding Officer of a battalion (Field Officer); Major Field Officer: rank held by second in command of a battalion; Captain: officer commanding a company; Lieutenant: junior company officer and Ensign (or Second Lieutenant): the most junior rank of officer, who often carried the regimental or King's colour into battle. (In the cavalry the rank of Ensign was replaced by Cornet).

The non-commissioned ranks in the infantry were (in descending order of seniority):

Sergeant Major, Quartermaster Sergeant, Sergeant, Corporal, Drummer and Private (in the Rifle Brigade Rifleman). The additional rank of Colour Sergeant was introduced in 1813 as a way of honouring deserving Sergeants, there being one Colour Sergeant appointed to each company.

In the cavalry the ranks were: Sergeant Major, Troop Quartermaster, Sergeant, Corporal, Trumpeter and Private (sometimes Trooper). In 1809 the Troop Quartermasters were replaced by Troop Sergeant Majors, who ranked immediately below the (regimental) Sergeant Major, and Quartermasters (who were commissioned officers) were introduced at the same time.

In the Household Cavalry (the 1 and 2 Regiments of Life Guards and, in 1827, the Royal Horse Guards) there were only two non-commissioned ranks: Quartermaster and Corporal. The rank of (regimental) Corporal Major

There are a number of groups of re-enactors who specialise in military themes. Seeing one of their displays can be one way to understand your ancestor's experiences in the Army.

was introduced in 1804. In the Household Cavalry, Corporals equated with Sergeants in the remainder of the Army.

Ranks in the Artillery were similar, but with additional categories, bearing in mind that guns had to be transported, and this was originally done mostly with horses, which needed to be fed, watered, equipped and stabled. The ranks, in descending order from the troop commander, would be:

Captain, Second Captain, Lieutenant, Staff Sergeant, Sergeant, Corporal, Bombardier, Gunner, Driver, Farrier, Shoeing Smith, Collar Maker, and Wheelwright.

Adapted from a table which appeared in *Ancestors Magazine* in April 2004.

Index